T0206078

Lecture Notes in Computer Science   **14510**

Founding Editors

Gerhard Goos
Juris Hartmanis

Editorial Board Members

Elisa Bertino, *Purdue University, West Lafayette, IN, USA*
Wen Gao, *Peking University, Beijing, China*
Bernhard Steffen ⓘ, *TU Dortmund University, Dortmund, Germany*
Moti Yung ⓘ, *Columbia University, New York, NY, USA*

The series Lecture Notes in Computer Science (LNCS), including its subseries Lecture Notes in Artificial Intelligence (LNAI) and Lecture Notes in Bioinformatics (LNBI), has established itself as a medium for the publication of new developments in computer science and information technology research, teaching, and education.

LNCS enjoys close cooperation with the computer science R & D community, the series counts many renowned academics among its volume editors and paper authors, and collaborates with prestigious societies. Its mission is to serve this international community by providing an invaluable service, mainly focused on the publication of conference and workshop proceedings and postproceedings. LNCS commenced publication in 1973.

Jaideep Vaidya · Moncef Gabbouj · Jin Li
Editors

Artificial Intelligence Security and Privacy

First International Conference
on Artificial Intelligence Security and Privacy, AIS&P 2023
Guangzhou, China, December 3–5, 2023
Proceedings, Part II

 Springer

Editors
Jaideep Vaidya
Rutgers University
Newark, NJ, USA

Moncef Gabbouj
Tampere University
Tampere, Finland

Jin Li
Guangzhou University
Guangzhou, China

ISSN 0302-9743   ISSN 1611-3349 (electronic)
Lecture Notes in Computer Science
ISBN 978-981-99-9787-9   ISBN 978-981-99-9788-6 (eBook)
https://doi.org/10.1007/978-981-99-9788-6

This Springer imprint is published by the registered company Springer Nature Singapore Pte Ltd.
The registered company address is: 152 Beach Road, #21-01/04 Gateway East, Singapore 189721, Singapore

Paper in this product is recyclable.

Preface

The first International Conference on Artificial Intelligence Security and Privacy (AIS&P 2023) was held in Guangzhou, China during December 3–5, 2023. AIS&P serves as an international conferences for researchers to exchange the latest research progress in all areas such as artificial intelligence, security and privacy, and their applications. This volume contains papers presented at AIS&P 2023.

The conference received 115 submissions. The committee accepted 40 regular papers and 23 workshop papers to be included in the conference program. Every paper received 2 or 3 Single-blind reviews. These proceedings contain revised versions of the accepted papers. While revisions were expected to take the referees' comments into account, this was not enforced and the authors bear full responsibility for the content of their papers.

AIS&P 2023 was organized by Huangpu Research School of Guangzhou University. The conference would not have been such a success without the support of these organizations, and we sincerely thank them for their continued assistance and support.

We would also like to thank the authors who submitted their papers to AIS&P 2023, and the conference attendees for their interest and support. We thank the Organizing Committee for their time and effort dedicated to arranging the conference. This allowed us to focus on the paper selection and deal with the scientific program. We thank the Program Committee members and the external reviewers for their hard work in reviewing the submissions; the conference would not have been possible without their expert reviews. Finally, we thank the EasyChair system and its operators, for making the entire process of managing the conference convenient.

November 2023

Chunsheng Yang
Haibo Hu
Changyu Dong

Organization

General Chairs

Chunsheng Yang   National Research Council, Canada
Haibo Hu   Hong Kong Polytechnic University, China
Changyu Dong   Guangzhou University, China

Program Chairs

Jaideep Vaidya   Rutgers University, USA
Moncef Gabbouj   Tampere University, Finland
Jin Li   Guangzhou University, China

Track Chairs

Muhammad Khurram Khan   King Saud University, Saudi Arabia
Yun Peng   Guangzhou University, China
Kwangjo Kim   KAIST, South Korea
Shaowei Wang   Guangzhou University, China

Publication Chairs

Weizhi Meng   Technical University of Denmark, Denmark
Francesco Palmieri   University of Salerno, Italy

Publicity Chairs

Hongyang Yan   Guangzhou University, China
Yu Wang   Guangzhou University, China

Steering Committee

Albert Zomaya	University of Sydney, Australia
Jaideep Vaidya	Rutgers University, USA
Moncef Gabbouj	Tampere University, Finland
Jin Li	Guangzhou University, China

Contents – Part II

Contents – Part I

Application of Lattice-Based Unique Ring Signature in Blockchain Transactions

Fengyin Li(✉) ⓘ, Junhui Wang, Dandan Zhang, Guoping Li, and Xilong Yu

Qufu Normal University, Rizhao, China
lfyin318@qfnu.edu.cn

Abstract. Ring signature schemes provide anonymity but suffer from the double-spending problem if uncontrolled overused in blockchain transactions. In the unique ring signature scheme, each member in the ring can only generate one signature for one message at most on behalf of the ring, which can effectively prevent the problem of double spending. However, most of the existing unique ring signature schemes are constructed based on the difficult problems of traditional cryptography, which are not safe in the post-quantum environment. This paper design and propose a lattice based unique ring signature scheme, which can provide anonymity protection for the transaction initiator when users trade, and can prevent malicious users from double flower attacks, ensuring the legitimacy of the transaction, and this scheme is safe in the quantum environment.

Keywords: lattice · quantum resistance · unique ring signature · blockchain · privacy protection

1 Introduction

Ring signature was first proposed by Rivest, Shamir and Tauman [1], aiming at realizing anonymous technology. In a ring signature scheme, only the signer himself but no one else knows who the real signer is. The ring is automatically and dynamically generated by the signer, who can use his own private key and other members' public keys to generate signatures without the knowledge of other members. The verifier uses the public key set of all members in the ring to verify the validity of the signature, and the verifier can only confirm that the signature comes from the changed ring, but cannot confirm which member generated the signature. That is, the verifier cannot trace the real identity of the signer, therefore, the ring signature scheme provides anonymity. However, the strong anonymity of ring signature schemes can be uncontrollably overused in some situations, such as in the double-spending problem of cryptocurrencies [2].

Franklin and Zhang [3, 4] introduced a unique ring signature. There is a unique label in the unique ring signature to ensure the uniqueness of the signature. The uniqueness ensures that any signer in the same ring can only sign the same message once valid signature. In addition to uniqueness, the unique ring signature scheme also provides anonymity and unforgeability. The unique ring signature scheme can be applied

© The Author(s), under exclusive license to Springer Nature Singapore Pte Ltd. 2024
J. Vaidya et al. (Eds.): AIS&P 2023, LNCS 14510, pp. 1–9, 2024.
https://doi.org/10.1007/978-981-99-9788-6_1

to scenarios such as electronic voting systems, electronic token systems, and electronic currency transactions.

In 2012, Franklin and Zhang [3, 4] first proposed a unique ring signature scheme, and the following year improved the unique ring signature scheme. In 2016, Rebekah applied the unique ring signature scheme to blockchain and analyzed that in some blockchain applications, the unique ring signature is superior to the linkable ring signature [5]. In 2021, Ta et al. improved the signature size of a unique ring signature and proposed a unique ring signature scheme [6] with a signature size smaller than schemes [3] to [5]. With the development of quantum computers, the security of signature schemes based on traditional difficult problems will be threatened. In the field of post quantum cryptography, the main motivation of researchers is to construct security methods to resist potential threats from quantum computers. In this regard, lattice based cryptography has attracted widespread attention due to its high efficiency, simplicity, high parallelism and strong provable security.

2   Preliminaries

2.1   Lattice

Definition 1 (Lattice). Lattice [7] is the set of n linear combinations of all integer coefficients of $\Lambda = L(x_1, x_2, \ldots, x_n) = \left\{ \sum_{i=1}^{n} a_i b_i | a_i \in z \right\}$ linearly independent vector groups, namely: $x_1, x_2, \ldots, x_n$.

Definition 2 (Full-rank lattice). Define m-dimensional full-rank q-ary lattice [8] as: $\Lambda_q^{\perp}(A) = \{x \in Z^m | Ax = 0(\bmod q)\}$, $\Lambda_q^u(A) = \{x \in Z^m | Ax = u(\bmod q)\}$. Among them, q is a prime number, m and n are positive integers, matrices $A \in Z_q^{n \times m}$, and vectors $u \in Z_q^n$. $\Lambda_q^{\perp}(A)$ and $\Lambda_q^u(A)$ can be abbreviated as $\Lambda^{\perp}(A)$ and $\Lambda^u(A)$.

Definition 3 (SIS problem). Given an integer q, a matrix $A \in Z_q^{n \times m}$ and a real number β, find a non-zero vector e such that $A \cdot e = 0 \bmod q$, $0 < \|e\| \leq \beta$ and such a problem [8] is called SIS problem.

Definition 4 (LWE problem). Let q and m be functions of n, where $q > 2$, χ is a discrete normal error distribution, and the LWE problem [9] involves finding a vector s and an error vector e in a lattice, such that for a given public key matrix A, there is $A \cdot s + e = b \bmod q$, where b is the objective vector.

Definition 5 (Decision LWR). For a vector $s \in Z_q^n$, define the LWR distribution L_s to be the distribuiton over $Z_q^n \times Z_p$ obtained by choosing a vector $a \leftarrow Z_q^n$ and outputting $(a, b = \lfloor \langle a \cdot s \rangle \rfloor_p$. The decision LWR [10] is to distinguish between m independent samples $(a_i, b_i) \leftarrow L_s$, and m samples drawn uniformly and independently from $Z_q^n \times Z_p$.

2.2 Security Modele

The unique ring signature scheme provides anonymity, unforgeability and uniqueness, and its security models are as follows:

Anonymity. Parameter Generation: Challenger C runs URS.Setup(1^λ) to obtain the public parameter pp when entering security parameters λ. Then challenger C runs URS.KeyGen(pp) to get the key pair $(pk_i, sk_i) \leftarrow$ URS.KeyGen(pp) for each user i $(i = 1, 2, \ldots, N)$. C selects the public key set $S = \{pk_i\}_{i=1}^{N}$, and sends (pp,S) to adversary A.

Query: Adversary A can conduct a polynomially limited number of private key query and signature query to the random oracle.

Key query: A can query the private key of any member i in the ring, and C sends the private key sk_i to A.

Signature query: A can ask any member i in the ring to sign the message m, and C will σ_i send the signature to A.

Forgery: A selects two numbers i_0, i_1 and a message m^* and a ring $R^*(R^* \subseteq S)$, sends the tuple (i_0, i_1, m^*, R^*) to C, C randomly and uniformly selects a bit b $\leftarrow \{0,1\}$, runs $\sigma \leftarrow$ URS.sign(m^*, R^*, sk_{i_b}). C returns σ to A.

If b ' = b, A will win the game, where A's advantage in winning is:

$$\text{Adv}_{A,C}^{Anonymity}(\lambda) := \Pr[b' = b] - 1/2 = \text{negl}(\lambda)$$

Unforgeability. Parameter Generation: Challenger C runs URS.Setup(1^λ) to obtain the public parameter pp when entering security parameters λ. Then challenger C runs URS.KeyGen(pp) to get the key pair $(pk_i, sk_i) \leftarrow$ URS.KeyGen(pp) for each user i $(i = 1, 2, \ldots, N)$. C selects the public key set $S = \{pk_i\}_{i=1}^{N}$, and sends (pp,S) to adversary A.

Query: Adversary A can conduct a polynomially limited number of private key query and signature query to the random oracle.

Key query: A can query the private key of any member i in the ring, and C sends the private key sk_i to A.

Signature query: A can ask any member i in the ring to sign the message m, and C will σ_i send the signature to A.

Forgery: Attacker A outputs a ring signature on a σ^* message m^* and a ring $R^*(R^* \subseteq S)$, and A has never asked for a signature on $(\forall, R^*, m^*)$ before.

If URS.$Verify(\sigma^*, R^*, \sigma^*) = 1$, A will win the game, and the advantage of A winning is:

$$\text{Adv}_{A,C}^{Unforge}(\lambda) := \Pr[A \text{ wins}] = \text{negl}(\lambda)$$

Uniqueness. Parameter Generation: Challenger C runs URS. Setup(1^λ) to obtain the public parameter pp when entering security parameters λ. Then challenger C runs URS. KeyGen(pp) to get the key pair $(pk_i, sk_i) \leftarrow$ URS.KeyGen(pp) for each user i $(i = 1, 2, \ldots, N)$. C selects the public key set $S = \{pk_i\}_{i=1}^{N}$, and sends (pp,S) to adversary A.

Query: Adversary A can conduct a polynomially limited number of private key query and signature query to the random oracle.

Key query: A can query the private key of any member i in the ring, and C sends the private key sk_i to A.

Signature query: A can ask any member i in the ring to sign the message m, and C will σ_i send the signature to A.

Forgery: Attacker A outputs valid t signatures $\sigma_1, \sigma_2, \ldots, \sigma_t$ for the same message m^* for the same ring R^* that are different. The challenger parses the signature as $\sigma_j = (\tau_j, \pi_j)$, and checks whether there is a unique label τ_k, and whether $(k = 1, 2, \ldots, t)$ is pairwise different.

If the unique labels are different, A will win the game, where the advantage of A winning is:

$$\text{Adv}_{A,C}^{Unique}(\lambda) := \Pr[A\ wins] = \text{negl}(\lambda)$$

3   Preliminaries

3.1   Framework of the Scheme

A Unique Ring Signature (URS) scheme consists of four algorithms: parameter generation, key generation, signature, verification. URS = (URS.Setup, URS.KeyGen, URS.Sign, URS.Verify).

URS.Setup(1^λ): This probabilistic polynomial time (PPT) algorithm takes security parameters λ as input and outputs public parameters pp.

URS.KeyGen(pp): This algorithm takes a public parameter pp as input to generate a signature private key sk and a public key pk. This algorithm will generate a key pair for each user.

URS.Sign(pp,m,R,sk): The PPT algorithm outputs a signature σ on the message m, the ring $R = (pk_1, pk_2, \ldots, pk_N)$, for URS, the signature σ can be parsed as $\sigma = (\tau, \pi)$, where τ is called the unique label.

URS.Verify(pp,m,R,σ): This deterministic polynomial time (DPT) algorithm takes as input the public parameter pp, the message m, the signer ring R, and the ring signature σ, and returns 1 if the signature σ is valid, Otherwise return 0.

3.2   Security of the Scheme

The unique ring signature scheme provides anonymity, unforgeability and uniqueness, and its security proofs are as follows:

Anonymity. Under the Random oracle model, according to the difficulty assumption of LWE problem, the unique ring signature scheme on lattice has anonymity. Assuming the existence of a PPT adversary A, after polynomial degree questioning, it is possible to distinguish between two legitimate signatures with an undeniable probability, breaking the anonymity of the given lattice unique ring signature scheme. Therefore, we can construct a simulator C with non-negligible advantages, which can solve examples of LWE problems.

Unforgeability. Under the Random oracle model, according to the difficulty assumption of the SIS problem, the unique ring signature scheme on the lattice is unforgeable. Assuming the existence of a PPT adversary A, who can forge a signature with a non-negligible degree after polynomial questioning, the non-forgeability of the given lattice unique ring signature scheme is compromised. Therefore, we can construct a simulator C with an undeniable advantage, which can solve an instance of SIS problems.

Uniqueness. Under the Random oracle model, according to the difficulty assumption of LWR problem, the unique ring signature scheme on lattice is unique. Assuming the existence of a PPT adversary A, who can forge a signature with existing information in a non-negligible manner after polynomial level questioning, the uniqueness of the given lattice unique ring signature scheme is compromised. Therefore, we can construct a simulator C with an undeniable advantage, which can solve an example of the LWR problem.

4 Blockchain Transaction Based on Unique Ring Signature Scheme on Lattice

Blockchain is a decentralized storage structure. Using blockchain to store data has the characteristics of safety, reliability, non-repudiation and non-tampering. These characteristics make blockchain technology widely used in finance, medical care, etc. prospect. However, since the ledger information for recording transactions in the blockchain is public, each participant can access all transaction data, which leads to the leakage of user privacy and restricts the development of blockchain applications [11].

4.1 Blockchain Privacy Threats and Countermeasures

The privacy threats faced by blockchain mainly include identity privacy threats and data privacy threats [12]. Identity privacy threat refers to attackers obtaining the identity information of traders by combining some background knowledge based on analyzing transaction data. Data privacy threat refers to the ability of attackers to obtain valuable information through analyzing transaction records, such as the flow of funds in a certain account, transaction details, and other information.

There are two main types of identity privacy protection schemes [13]: mixed currency schemes and schemes based on cryptography technology. The mixed currency scheme is divided into centralized mixed currency scheme and decentralized mixed currency scheme, and schemes based on cryptography technology have zero knowledge proof and ring signature. Data privacy protection schemes mainly include homomorphic encryption algorithms, searchable encryption algorithms, and attribute based encryption algorithms.

In order to protect the privacy of users, the ring signature technology can be used to realize the anonymity of the transaction initiator. When the user trades, the transaction initiator can independently form a ring pair transaction signature with the user on the chain, so that the transaction cannot be traced, thus realizing the user privacy protection. However, once the ring signature technology is exploited by malicious users, double spending attacks will occur. Using the unique ring signature scheme proposed in this paper, if a user uses the same ring to sign the same message twice, the two signatures of the user will be linked, and one of the signatures will be invalid. In this way, the anonymity of the ring signature technology can not only be used to hide the identity address of the transaction initiator, but also ensure the legality of the signature.

When conducting a transaction, the initiator of the transaction independently selects some member rings, and then signs the transaction using their own private key. A uniqueness label is attached to the signature to prove that the signature was signed by a user using ring R on message m. The verifier first verifies the label, and if the label is linked to the label in the published signature, it indicates that this expenditure has been spent twice and is an illegal transaction. If the tag cannot be linked to the tag in the publicly announced signature, it proves that the transaction is unique, that is, the money has not been spent repeatedly. Then, the verifier uses the public key set of all members in the ring to verify the validity of the signature. If the signature verification is successful, the transaction can proceed. If the verification is not successful, the transaction is illegal. The entire process of the transaction is shown in Fig. 1.

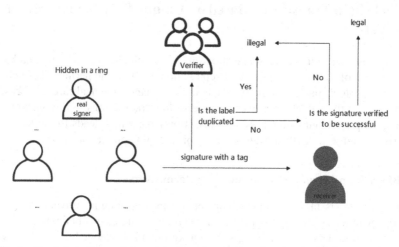

Fig. 1. Blockchain transaction process based on unique ring signature scheme on lattice

4.2 Privacy Protection for Blockchain Transactions Based on Lattice UniquRing Signature Scheme

When conducting transactions in blockchain, the sender will transfer funds to the recipient's public key address, which is obtained by hashing the public key generated by the recipient's private key. The process of generating a public key address is unidirectional and irreversible. The address cannot reverse the public key, and the public key cannot reverse the private key. Due to this characteristic, public key addresses can be publicly used on the blockchain, and anyone can obtain someone else's public key address, but the corresponding private key cannot be inferred.

Generate Public and Private Keys and Addresses. When conducting transactions in blockchain, the sender will transfer funds to the recipient's public key address, which is obtained by hashing the public key generated by the recipient's private key. The process of generating a public key address is unidirectional and irreversible. The address cannot reverse the public key, and the public key cannot reverse the private key. Due to this characteristic, public key addresses can be publicly used on the blockchain, and anyone can obtain someone else's public key address, but the corresponding private key cannot be inferred.

The condition for user transactions is to have a public key address that can accept transfer transactions and a private key that can sign transactions. The user performs the following actions to generate a private key and address:

The user U_i calls the Setup algorithm and the URS.KenGen algorithm, generates a private key SK_{U_i}, and then calculates $PK_{U_i} = A \cdot SK_{U_i}$ as a public key, thus U_i having a public private key pair (SK_{U_i}, PK_{U_i}).

Then U_i use the hash function to use the hash result of the public key $H(PK_{U_i})$ as the public key address, which will be publicly available in the blockchain, and the private key will store itself.

The same user and the same private key can control multiple public key addresses, and U_i can randomly select several random parameters $A_2, A_3, \ldots, A_n$ etc., which are the same distributed with A, and then calculate $PK_{2U_i} = A_2 \cdot SK_{2U_i}, PK_{3U_i} = A_3 \cdot SK_{3U_i}, \ldots, PK_{nU_i} = A_n \cdot SK_{nU_i}$.

Publish the hash results $H(PK_{2U_i}), H(PK_{3U_i}), \ldots, H(PK_{nU_i})$ of the derived public key as public key addresses.

Transaction. After possessing the public key address and private key, users can participate in blockchain transactions. This section applies the lattice based unique ring signature scheme proposed in the previous section to the blockchain transaction process to achieve a transaction scheme that provides privacy protection for the trading party. The process of transfer transactions between the sender U_i and the receiver U_k is as follows:

U_k spontaneously select some users' public key addresses as rings to hide themselves, and then use their own private key SK_{U_i} to sign the transaction. There will be a uniqueness label in the signature to prove that the signature was signed by a user using ring R on message M. The framework diagram for generating ring signatures in blockchain transactions is shown in Fig. 2.

Firstly, U_i verify the label. If the label is linked to a label in an existing signature, it indicates that message M has been signed using ring R. In order to achieve higher anonymity, it is necessary to replace the ring member before signing. If the tag is not linked to a tag in an existing signature, the transaction can be broadcasted. Subsequently, the verifier uses the public key set of all members in the ring to verify the validity of the signature.

If the signature verification passes, the transaction is legal, otherwise it is illegal. If this transaction is legal, U_k will be used as part of the transaction as the sender to verify the legitimacy of the transaction source. The blockchain transaction process framework using lattice based unique ring signature scheme is shown in Fig. 3.

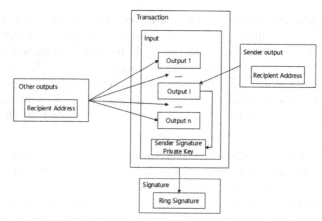

Fig. 2. Transaction sender generates a ring signature framework diagram in the transaction

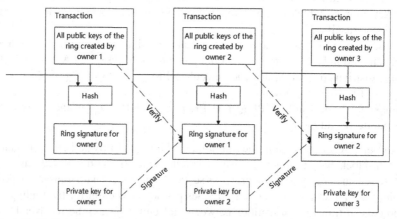

Fig. 3. A Blockchain Transaction Process Framework Using Lattice Based Unique Ring Signature Scheme

5 Summary

Unique ring signature scheme can effectively prevent double spending problems in blockchain transactions because each member in the ring can only generate one signature for one message on behalf of the ring. Most of the existing unique ring signature schemes are constructed based on the difficult problems of traditional cryptography, which are not safe in the post-quantum environment. This paper constructs a quantum-resistant ring signature scheme based on lattice, and applies it to blockchain transactions, which can ensure the legitimacy of transactions while protecting the privacy of transaction initiators.

References

1. Rivest, R.L., Shamir, A., Tauman, Y.: How to leak a secret. In: Boyd, C. (ed.) ASIACRYPT 2001. LNCS, vol. 2248, pp. 552–565. Springer, Heidelberg (2001). https://doi.org/10.1007/3-540-45682-1_32
2. Nakamoto, S.: Bitcoin: a peer-to-peer electronic cash system (2008)
3. Franklin, M., Zhang, H.: A framework for unique ring signatures.IACR Cryptol. ePrint Arch. **2012**, 577 (2012)
4. Franklin, M., Zhang, H.: Unique ring signatures: a practical construction. In: Sadeghi, A.-R. (ed.) FC 2013. LNCS, vol. 7859, pp. 162–170. Springer, Heidelberg (2013). https://doi.org/10.1007/978-3-642-39884-1_13
5. Mercer, R.: Privacy on the blockchain: Unique ring signatures. arXiv preprint arXiv:1612.01188 (2016)
6. Ta, A.T., et al: Efficient unique ring signature for blockchain privacy protection. In: Baek, J., Ruj, S. (eds.) ACISP 2021. LNCS, vol. 13083, pp. 391–407. Springer, Cham (2021). https://doi.org/10.1007/978-3-030-90567-5_20
7. Regev, O.: Lattice-based cryptography. In: Dwork, C. (ed.) CRYPTO 2006. LNCS, vol. 4117, pp. 131–141. Springer, Heidelberg (2006). https://doi.org/10.1007/11818175_8
8. Gentry, C., Peikert, C, Vaikuntanathan, V.: Trapdoors for hard lattices and new cryptographic constructions. In: Proceedings of the Fortieth Annual ACM Symposium on Theory of Computing, pp. 197–206 (2008)
9. Rivest, R.L., Shamir, A., Tauman, Y.: How to leak a secret. In: Boyd, C. (ed.) ASIACRYPT 2001. LNCS, vol. 2248, pp. 552–565. Springer, Heidelberg (2001). https://doi.org/10.1007/3-540-45682-1_32
10. Banerjee, A., Peikert, C., Rosen, A.: Pseudorandom functions and lattices. In: Pointcheval, D., Johansson, T. (eds.) EUROCRYPT 2012. LNCS, vol. 7237, pp. 719–737. Springer, Heidelberg (2012). https://doi.org/10.1007/978-3-642-29011-4_42
11. Duan, J., Gu, L., Zheng, S.: Polymerized RingCT: an efficient linkable ring signature for ring confidential transactions in blockchain. J. Phys: Conf. Ser. **1738**(1), 012109 (2021)
12. Shen, C.: Research on blockchain privacy threats and protection technologies. Netw. Secur. Data Gov. **42**(04), 1–8 (2023)
13. Xie, Q., Yang, N., Feng, X.: Overview of privacy protection technologies for blockchain transactions. Comput. Appl. 1–14 (2023)

Rethinking Distribution Alignment for Inter-class Fairness

Jinhuang Ye[1], Jiawei Wu[2], Zuoyong Li[3](✉), and Xianghan Zheng[1](✉)

[1] College of Computer and Big Data, Fuzhou University, Fuzhou 350108, China
`xianghan.zheng@fzu.edu.cn`
[2] School of Intelligent Systems Engineering, Sun Yat-sen University, Shenzhen 518107, China
[3] Fujian Provincial Key Laboratory of Information Processing and Intelligent Control, College of Computer and Control Engineering, Minjiang University, Fuzhou 350121, China
`fzulzytdq@126.com`

Abstract. Semi-supervised learning (SSL) is a successful paradigm that can use unlabelled data to alleviate the labelling cost problem in supervised learning. However, the excellent performance brought by SSL does not transfer well to the task of class imbalance. The reason is that the class bias of pseudo-labelling further misleads the decision boundary. To solve this problem, we propose a new plug-and-play approach to handle the class imbalance problem based on a theoretical extension and analysis of distribution alignment. The method, called Basis Transformation Based Distribution Alignment (BTDA), efficiently aligns class distributions while taking into account inter-class relationships.BTDA implements the basis transformation through a learnable transfer matrix, thereby reducing the performance loss caused by pseudo-labelling biases. Extensive experiments show that our proposed BTDA approach can significantly improve performance in class imbalance tasks in terms of both accuracy and recall metrics when integrated with advanced SSL algorithms. Although the idea of BTDA is not complex, it can show advanced performance on datasets such as CIFAR and SVHN.

Keywords: Image classification · Semi-supervised learning · Distribution alignment · Basis transformation · Class-imbalanced datasets · Inter-class bias

1 Introduction

Semi-supervised learning (SSL) is a promising learning paradigm that can solve the problem of training models in the absence of large-scale labelled training data. SSL methods have become a hot research topic in various tasks due to the high cost of acquiring labelled data. Classical SSL methods have been developed with the aim of alleviating the burden of expensive labelling. These methods show significant performance on category-balanced datasets, in some cases even outperforming fully supervised learning, as shown in Fig. 1 (a)(c).

© The Author(s), under exclusive license to Springer Nature Singapore Pte Ltd. 2024
J. Vaidya et al. (Eds.): AIS&P 2023, LNCS 14510, pp. 10–21, 2024.
https://doi.org/10.1007/978-981-99-9788-6_2

However, the performance of semi-supervised algorithms can be greatly degraded when they encounter the imbalanced data commonly found in life. As shown in the Fig. 1, even SSL algorithms considered state-of-the-art in the balanced case can suffer up to 40% performance degradation when faced with unknown distributions. This is due to the fact that pseudo-labelling introduces a strong category bias, leading to misguided optimisation and irreversible consequences. To alleviate this problem and to maintain the performance of semi-supervised learning in the presence of class imbalance, we further analyse the algorithm with the best performance maintenance in our experiments.

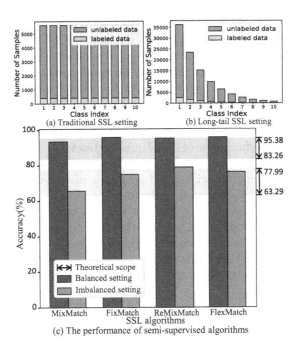

Fig. 1. Experimental results in different Settings. The performance of semi-supervised algorithms (c) approach the theoretical limitation (i.e., the performance of fully-supervised learning in entire dataset) under the balanced setting (a), while degrading significantly under the imbalanced setting (b). Notably, imbalanced SSL algorithm, ReMixMatch [1], surpasses the theoretical limitation. Hence, the question naturally arise: How can imbalanced SSL algorithms maintain their performance in class-imbalance datasets?

Unlike the standard SSL algorithm (MixMatch [2]), ReMixMatch [1] integrates distribution alignment (DA) and rotation loss, while improving the traditional components of SSL (e.g., data enhancement and consistency loss). Therefore, we perform an ablation analysis of DA and rotation loss under different degrees of class imbalance. Figure 2 shows that both DA and rotation loss can effectively solve the class imbalance problem. While the SSL algorithm with DA removed shows a rapid performance degradation in the face of scenarios with

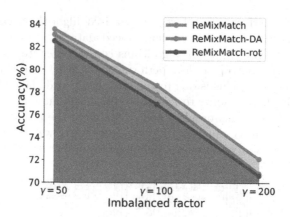

Fig. 2. Ablation results (Accuracy ↓) with ReMixMatch on CIFAR10-LT. On the X-axis, three different imbalanced factors γ are 50, 100 and 200 respectively.

increasing imbalance rates, it is evident that the role of DA is more important in maintaining the performance of imbalanced datasets.

By extending the theory of distribution alignment on linear algebra, we further optimize the original algorithm and propose a new component called Basis Transformation based Distribution Alignment (BTDA)[1]. Specifically, BTDA explicitly aligns the predicted distribution with its corresponding true distribution by learning a transition matrix. In addition, to obtain the optimal neural network and transition matrix, we develop a highly efficient closed-form optimization scheme during the SSL training.

To summarize, our contributions can be outlined as follows:

- Our analysis of existing DA techniques reveals that it is a special instance of basis transformation, providing theoretical guarantees for further improvement.
- We propose a plug-and-play component for existing SSL algorithms, called Basis Transformation based Distribution Alignment (BTDA). This method effectively mitigates the influence of pseudo-label bias by implicitly aligning the class distributions between unlabeled predictions and true distribution.
- Our comprehensive experimentation reveals that BTDA yields a significant improvement over baseline algorithm.

2   Related Work

2.1   Semi-Supervised Learning

With the continuous development of deep learning techniques, semi-supervised learning (SSL) as an emerging learning method has received a lot of attention

---

[1] Codes are available at https://github.com/211027128/BTDA.

from academia. Compared to traditional supervised learning, SSL has unparalleled potential in terms of cost savings. It is able to train a generic model with minimal manual annotation, thus significantly reducing the cost of labelling data. Consistency regularisation [13,14] and self-training [15,16] are two commonly used methods in SSL. Consistency regularisation requires the model to maintain a consistent output for perturbed inputs, thus improving the generalisation ability of the model. Self-training, on the other hand, assigns pseudo-labels to unlabelled data to compensate for the shortcomings of unlabelled data so that it can be fully exploited. To exploit the strengths of both techniques, some hybrid methods (e.g. ReMixMatch [1] and FlexMtach [4]) have taken semi-supervised learning to new heights of performance by combining both techniques.

2.2   Imbalanced Supervised Learning

The long-tail distribution [17] is a common data phenomenon in real life, where the number of samples in a few categories is much smaller than the number of samples in the majority category. This unbalanced data distribution can negatively affect the performance of supervised learning, as the model will prefer to predict the majority category and ignore the minority category. Common supervised learning methods for unbalanced tasks include class re-balancing [18,19], information augmentation [20,21] and model improvement [22,23]. However, the performance of these methods when applied to the SSL task is inconsistent due to the presence of unlabelled data with unknown distribution. Therefore, the problem of class-imbalanced semi-supervised learning remains a challenge. Our research aims to explore the potential of applying the ideas of supervised learning to semi-supervised learning algorithms, and to propose new methods inspired by existing methods to effectively solve the class imbalance problem in a semi-supervised setting.

2.3   Imbalanced Semi-supervised Learning

The class-imbalanced semi-supervised learning task is challenging because the bias in the amount of data can severely affect the performance of models trained by semi-supervised learning. Several approaches have been proposed to address this problem. One approach is Suppressed Consistency Loss (SCL) [5], which reduces the impact of a few class misclassifications on the model by suppressing consistency loss. Another approach, BI-Sampling (Bis) [6] employs a differentiated sampling strategy for unlabeled data using classifiers and feature extractors. In addition, Adaptive Deep Supervised Hashing (ADSH) [7] proposes a class-based threshold training method to optimise the pseudo-label selection process. These methods address the class imbalance problem in SSL from different aspects and show good results, which can help solve the class imbalance problem and improve the performance and accuracy of machine learning models.

A comprehensive review of these methods in this section allows us to better understand the challenges faced by existing methods and provides important insights for future research in this area.

3 Methodology

3.1 Preliminary

Given a labeled set $\mathcal{D}_l = \{(x_l, y_l)\}_{i=1}^N$ and a unlabeled set $\mathcal{D}_u = \{x_u\}_{i=1}^M$, we assume that the sample count of each class in labeled set follow $N_1 \geq N_2 \geq \cdots \geq N_K$, in which K denotes the number of classes and the imbalanced factor equals to $\gamma_l = N_1/N_K$.

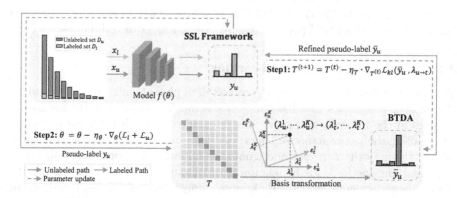

Fig. 3. Optimization of SSL with BTDA. The closed-form solution is obtained by two-step alternating optimization, in which the **step1** aims to optimize step $t+1$ for the transition matrix $\boldsymbol{T}^{(t+1)}$ under fixed $f(\theta)$ and the **step2** aims to optimize the neural network $f(\theta)$ under fixed T.

3.2 Distribution Alignment (DA)

Distribution alignment maximises the mutual information between input and output by aligning the predicted distribution with the true distribution. According to this principle, DA incorporates a form of fairness for pseudo-labels y_u, the position i of y_u proceeds as follow:

$$\bar{y}_u^i = y_u^i \times \frac{\tilde{q}^i(\mathcal{D}_l)}{\tilde{p}^i(\mathcal{D}_u)}, \tag{1}$$

where $\bar{y}_u, \tilde{q}(\mathcal{D}_l), \tilde{p}(\mathcal{D}_u)$ denote the refined pseudo-labels for supervision, the true distribution of unlabeled data (Equivalent to the true distribution of $\mathcal{D}_l$), and the distribution of predictions for unlabeled data (running average of predictions for unlabeled data during training), respectively.

Nonetheless, ReMixMatch's explication of the underlying principle merely elucidates the "what" aspect, while neglecting the "why" component. In order to understand the principle of DA starting, a mathematical derivation was followed by an explanation from a linear algebraic point of view. In linear algebra, a point in space has different eigenvalues in different basis vectors. Based on this theorem, we conclude that the distributions of labelled and unlabelled data are both constrained by the same space, but there is a virtual gap between them due to the different power of the supervised signal. Therefore, we assume $\tilde{q}(\mathcal{D}_l)$ and $\tilde{p}(\mathcal{D}_u)$ have different basis as follows:

$$
\begin{cases}
\tilde{q}(\mathcal{D}_l) = \begin{bmatrix} \varepsilon_t^1 \cdots \varepsilon_t^K \end{bmatrix} \begin{bmatrix} \lambda_t^1 \\ \vdots \\ \lambda_t^K \end{bmatrix} \\
\tilde{p}(\mathcal{D}_u) = \begin{bmatrix} \varepsilon_u^1 \cdots \varepsilon_u^K \end{bmatrix} \begin{bmatrix} \lambda_u^1 \\ \vdots \\ \lambda_u^K \end{bmatrix},
\end{cases}
\tag{2}
$$

Table 1. Comparison results (Accuracy ↓) on CIFAR10-LT, CIFAR100-LT and SVHN-LT. The best performance is shown in **bold**.

Database	CIFAR10-LT			SVHN-LT		CIFAR100-LT	
Imbalanced factor	$\gamma = 50$	$\gamma = 100$	$\gamma = 200$	$\gamma = 20$	$\gamma = 50$	$\gamma = 20$	$\gamma = 50$
MixMatch [2]	73.06%	65.54%	60.35%	–	–	–	–
FixMatch [3]	80.19%	74.53%	65.47%	–	–	–	–
FlexMatch [4]	81.33%	76.40%	68.77%	–	–	–	–
ReMixMatch - rotation loss	82.51%	76.93%	70.64%	–	–	–	–
ReMixMatch - DA	83.08%	77.24%	70.81%	89.88%	82.60%	66.17%	**60.10%**
w/ DA [1]	83.54%	78.58%	72.11%	91.11%	87.86%	**66.82%**	59.78%
w/ BiS [6]	83.74%	78.17%	72.27%	90.85%	84.07%	66.73%	59.74%
w/ SCL [5]	83.39%	77.73%	72.16%	89.53%	81.86%	66.59%	59.93%
w/ ADSH [7]	82.30%	77.76%	69.77%	86.25%	78.57%	65.39%	58.74%
w/ BTDA	**84.75%**	**78.92%**	**73.47%**	**91.49%**	**88.34%**	66.79%	59.89%

where ε_t^*, ε_u^* denote a basis vector in their respective basis, and λ_t^i, λ_u^i denote the i_{th} eigenvalue on them. On the premise of finding the transition matrix T and its transpose T^{-1} of the basis transformation, the Eq. 1 of DA can be reformulated as

$$
\lambda_{u \to t}^i = T^{-1}\left[\lambda_u^i\right], \text{ s.t. }
\begin{cases}
\lambda_{u \to t}^i = \bar{y}_u^i \\
\lambda_u^i = y_u^i \\
\bar{y}_u^i = y_u^i \times \frac{\tilde{q}^i(\mathcal{D}_l)}{\tilde{p}^i(\mathcal{D}_u)} \\
\left[\tilde{p}^i(\mathcal{D}_u)\right] = T\left[\tilde{q}^i(\mathcal{D}_l)\right],
\end{cases}
\tag{3}
$$

where $\lambda_{u \to t}^i$ denote the eigenvalue of the transition from the unlabeled to the true labeled domain. Therefore, DA can be regarded as a single-element basis transformation that serves to align the class-wise basis. An intuitive idea then is to expand the single element matrix into a full transition matrix, which would be a better implementation of the basis transformation. This leads us to propose a new algorithm called Basis Transformation Based Distribution Alignment (BTDA).

3.3 Basis Transformation Based Distribution Alignment

In the above, DA can be thought of as a method of converting an unlabelled distribution into a labelled one, as follows:

$$
\begin{bmatrix} \lambda_{u \to t}^1 \\ \vdots \\ \lambda_{u \to t}^K \end{bmatrix} = \boldsymbol{T}^{-1} \begin{bmatrix} \lambda_u^1 \\ \vdots \\ \lambda_u^K \end{bmatrix}, \tag{4}
$$

where $\begin{bmatrix} \varepsilon_t^1 \dots \varepsilon_t^K \end{bmatrix} = \begin{bmatrix} \varepsilon_u^1 \dots \varepsilon_u^K \end{bmatrix} \boldsymbol{T}$. The basis vectors ε_t and ε_u cannot be found directly, so we can only solve for $\boldsymbol{T}$ using the pair of coordinates $(\lambda_u, \lambda_{u \to t})$. In order to obtain this pair of coordinates, we propose a sampling strategy. The method uses DA to generate the corresponding coordinates $\lambda_{u \to t}$ under the true base space for λ_u. Once the precise coordinate pairs are obtained, we can judiciously compute the transition matrix $\boldsymbol{T}$ via matrix operations to effectuate the distribution alignment of basis transformation.

Although we can derive the transition matrix $\boldsymbol{T}$, potentially conditioned on the need for an exact neural network $f(\theta)$, by Eq. 4 above. Therefore, finding the optimal $\boldsymbol{T}$ and the optimal neural network $f(\theta)$ are a chicken-and-egg problem. As shown in Fig. 3, we have skilfully solved this problem by introducing the idea of circular iteration.

Solving $\boldsymbol{T}$ Given a Fixed θ. Once the parameters θ of the neural network is fixed, we can solve for the transition matrix $\boldsymbol{T}$ according to Eq. 4. We propose to optimize step $t + 1$ for $\boldsymbol{T}^{(t+1)}$ by stochastic gradient descent (SGD), and use KL divergence as loss function $\mathcal{L}_{kl}$:

$$
\boldsymbol{T}^{(t+1)} = \boldsymbol{T}^{(t)} - \eta_T \nabla_{\boldsymbol{T}^{(t)}} \mathcal{L}_{kl}, \tag{5}
$$

where

$$
\mathcal{L}_{kl} = \lambda_{u \to t} \log \frac{\lambda_{u \to t}}{\boldsymbol{T}^{-1(t)} f(x_u; \theta)}, \tag{6}
$$

η_T is the update rate and $\nabla_{\boldsymbol{T}^{(t)}}$ represents the gradient of the transition matrix $\boldsymbol{T}$ at moment t.

Solving θ Given a Fixed $\boldsymbol{T}$. With the transition matrix $\boldsymbol{T}$ fixed, we can optimise the parameters θ of the neural network with smaller biased pseudo-labels. we optimize θ using the SGD algorithm, in which the unsupervised loss $\mathcal{L}_u$ is given by

$$
\mathcal{L}_u = \mathcal{H}(\boldsymbol{T}^{-1} y_u, f(x_u; \theta)), \tag{7}
$$

where $\mathcal{H}(\cdot, \cdot)$ denotes a distance metric, e.g., cross entropy. The refined pseudo-label $\boldsymbol{T}^{-1} y_u$ continuously optimises the neural network by cross-entropy in iterations, reducing the deviation of the two basis spaces. When the deviations are completely eliminated, we end up with a transition matrix with a unitary array.

With our proposed BTDA method, the neural network is made to correct the bias towards different categories. The increased emphasis on the tail class allows the neural network not to predict all the pseudo-labels as the head class, effectively dealing with the class imbalance problem.

4   Experiments

4.1   Experiments Settings

Imbalanced Datasets. All experiments are conducted with varying degrees of class-imbalance on CIFAR-10 [9], CIFAR-100, and SVHN [8]. CIFAR contains 60,000 32×32 colour images in 10 categories, 50,000 for training and 10,000 for testing, while SVHN contains 73,257 32×32 colour images in 10 categories, with 73,257 in the training set and 26,032 in the testing set. In addition, to meet experimental needs, we adjusted the degree of imbalance in the dataset using an imbalance factor γ. Specifically, the number of each class is set to $N_k = N_1 * \gamma_l^{k/(K-1)}$ and $M_k = M_1 * \gamma_u^{k/(K-1)}$.

Table 2. Comparison results (Recall ↓) of each class on CIFAR10-LT with $\gamma = 100$. Head classes: 1,2,3, medium classes: 4,5,6,7,8, tail classes: 9,10. The best performance is shown in **bold**.

SSL Methods	Class Index										Average
	1	2	3	4	5	6	7	8	9	10	
ReMixMatch-DA	99.2%	99.8%	92.7%	83.7%	87.9%	69.6%	79.4%	71.0%	51.5%	37.6%	77.24%
	97.23%			78.32%					44.55%		77.24%
w/ DA [1]	99.4%	99.6%	92.0%	84.1%	89.1%	69.4%	79.3%	70.2%	57.8%	38.8%	77.97%
	97%(-0.23%)			78.42%(+0.1%)					48.3%(+3.75%)		77.97%(+0.73%)
w/ BIS [6]	99.3%	99.9%	90.1%	86.8%	90.6%	69.7%	75.3%	71.7%	63.6%	39.4%	78.64%
	96.43%(-0.80%)			78.82%(+0.5%)					51.5%(+6.95%)		78.64%(+1.40%)
w/ SCL [5]	99.5%	99.7%	92.3%	83.7%	86.6%	70.9%	79.7%	69.2%	55.4%	40.3%	77.73%
	97.17%(-0.06%)			78.02%(-0.3%)					47.85%(+3.3%)		77.73%(+0.49%)
w/ ADSH [7]	98.5%	99.7%	94.5%	79.9%	87.1%	71.6%	79.1%	72.5%	60.1%	34.6%	77.76%
	97.56%(+0.33%)			78.04%(-0.28%)					47.35%(+2.8%)		77.76%(+0.52%)
w/ BTDA	99.1%	99.9%	93.1%	85.6%	89.0%	70.0%	79.3%	75.2%	62.9%	41.6%	79.57%
	97.36%(+0.13%)			**79.82%(+1.5%)**					**52.25%(+7.7%)**		**79.57%(+2.33%)**

Compared Methods. Our baseline is ReMixMatch-DA, which uses ReMix-Match [1] but without distribution alignment. The dominantly compared methods include four imbalanced semi-supervised methods, i.e., ReMixMatch [1], SCL [5], BIS [6] and ADSH [7].

Implementation Details. We refer to the pytorch codebase of FlexMatch [4][2] for all experiments. The WideResNet (WRN) [10] is used as the neural network for classification. During all the network training, we use stochastic gradient descent (SGD) with momentum 0.9 [11,12] as optimizer, iterating 100000 steps with batch size 64. The momentum of exponential moving average (EMA) for updating the distributions are all set with 0.999. To ensure a fair comparison, the same data enhancement method as the baseline algorithm was used for all experiments.

---

[2] https://github.com/TorchSSL/TorchSSL.

4.2   Experimental Results

Comparison Results of Accuracy. A series of experiments were conducted to compare the superior performance of BTDA over other algorithms. Firstly, a comparison with other SSL frameworks was made on three different sets of CIFAR-10LT with different imbalance factors, as presented in Table 1. Our results show that (i) ReMixMatch outperforms other SSL frameworks and that excluding DA leads to performance degradation. This demonstrates the superiority of DA in dealing with class imbalance problems. (ii) BTDA significantly improves the raw performance of ReMixMatch and demonstrates the potential to improve performance in different experimental environments. This reflects the fact that BTDA can better achieve distribution alignment and reduce inter-class bias.

To demonstrate the generality of BTDA, we compare it with the advanced Imbalanced Semi-Supervised Learning (ISSL) algorithm on a dataset with the addition of SVHN and CIFAR100. The results show that BTDA achieves the best results in all cases where there are few categories. When the number of categories is large, the difference in optimisation between the different algorithms is minimal. Therefore, the proposed BTDA approach can be considered as a promising solution to address the challenges of imbalanced data in real-world scenarios.

Comparison Results of Each Class. To investigate the specific impact of BTDA on long-tail classes, we conducted an additional comparative experiment, as presented in Table 2. The classes were categorized into head, medium, and tail classes. Our results show that BTDA has consistently improved recall for both the middle and tail categories, while maintaining recall for the head category. This includes a significant 7.7% improvement for the tail class, indicating a correction for class bias. BTDA is also the algorithm with the highest average recall improvement. These findings lend support to the effectiveness of BTDA in promoting fairness across classes, thereby highlighting its potential as a viable approach to addressing the long-tail problem in machine learning.

4.3   Detailed Analyses

Q1: Is DA Effective? To explore whether DA is effective, we investigated the KL divergence values between the unlabeled predicted $\tilde{p}(\mathcal{D}_u)$ and true distributions $p(\mathcal{D}_u)$ during machine learning model training and plotted the corresponding change curves. The experimental results presented in Fig. 4 provide evidence that distribution alignment can enhance the accuracy of the prediction distribution for predictions for unlabeled data by inducing the model to make non-uniform predictions for each category. It is worth noting that, for the purpose of enhancing the clarity of the KL divergence change pattern, we have made the decision to truncate the segment with significant divergence in the initial 10,000 iterations. The experimental results indicate that distribution alignment effectively reduces the KL divergence between probability distributions $\tilde{p}(\mathcal{D}_u)$ and $p(\mathcal{D}_u)$. Building upon the efficient mechanism of DA, we have proposed a novel approach called BTDA that offers further optimization. To explicate the underlying mechanisms and efficacy of BTDA, we present another detailed experiment.

Q2: Why is BTDA Effective? To circumvent the idiosyncrasies of BTDA for ReMixMatch, we investigate the rationale behind the effectiveness of BTDA by leveraging MixMatch [2]. As depicted in Fig. 5, MixMatch exhibits pronounced performance degradation in the tail classes. Owing to the misclassification of the tail classes as head classes, the accuracy of the tail classes is a mere $0.11 \sim 0.12$, only marginally superior to random guessing. By contrast, BTDA, equipped with a DA-based sampler, substantially enhances fairness, attaining accuracies of $0.26 \sim 0.40$ for the tail classes, while concurrently mitigating the probability of misclassification as the head classes. Consequently, BTDA augments performance by instilling fairness in the context of imbalanced SSL.

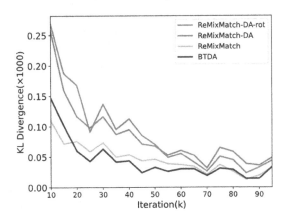

Fig. 4. KL divergence $\downarrow$ between the unlabeled predictive distribution $\tilde{p}(\mathcal{D}_u)$ and the true distribution $p(\mathcal{D}_u)$ in CIFAR10-LT dataset.

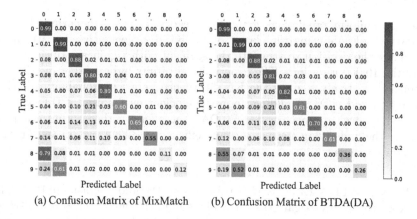

(a) Confusion Matrix of MixMatch   (b) Confusion Matrix of BTDA(DA)

Fig. 5. Confusion matrices on CIFAR10-LT under $\gamma_l = \gamma_u = 100$ using MixMatch [2]. BTDA significantly improves the performance of tail classes.

5   Conclusion

This work focuses on addressing the performance degradation of SSL algorithms caused by the class-imbalance problem. We propose a plug-and-play Basis Transformation based Distributed Alignment (BTDA) strategy for imbalanced SSL. BTDA extends the theory of distribution alignment on linear algebra, establishing class-to-class relationships for aligning basis. We also introduce alternating optimization to jointly learn a learnable transition matrix and neural network. This work significantly advances the theory of class-imbalanced semi-supervised tasks and provides a flexible component that can be integrated into existing SSL algorithms. Although we have taken the initial step, our plan is to further develop our BTDA approach by incorporating representation-level alignment. This will enable us to simultaneously fulfill both optimization conditions for ISSL in the future.

Acknowledgement. This research was supported by National Natural Science Foundation of China (61972187); Open Project of Key Laboratory of Medical Big Data Engineering in Fujian Province (KLKF202301); R&d Plan of Guangdong Province in key areas (2020B0101090005); the specific research fund of The Innovation Platform for Academician of Hainan Province (YSPTZX202145); Fujian Provincial Science and Technology Department Guided Project (2022H0012).

References

1. Berthelot, D., et al.: ReMixMatch: semi-supervised learning with distribution matching and augmentation anchoring. In: International Conference on Learning Representations (2020)
2. Berthelot, D., et al.: Mixmatch: a holistic approach to semi-supervised learning. Adv. Neural Inf. Process. Syst. **32** (2019)

3. Sohn, K., et al.: Fixmatch: simplifying semi-supervised learning with consistency and confidence. Adv. Neural Inf. Process. Syst. **33**, 596–608 (2020)
4. Zhang, B., et al.: Flexmatch: boosting semi-supervised learning with curriculum pseudo labeling. Adv. Neural Inf. Process. Syst. **34**, 18408–18419 (2021)
5. Hyun, M., et al.: Class-imbalanced semi-supervised learning. In: ICLR RobustML Workshop (2021)
6. He, J., et al.: Rethinking re-sampling in imbalanced semi-supervised learning. arXiv preprint arXiv:2106.00209 (2021)
7. Guo, L.-Z., Li, Y.F.: Class-imbalanced semi-supervised learning with adaptive thresholding. In: International Conference on Machine Learning. PMLR (2022)
8. Netzer, Y., et al.: Reading digits in natural images with unsupervised feature learning. In: NIPS Workshop on Deep Learning and Unsupervised Feature Learning (2011)
9. Krizhevsky, A., et al.: Learning multiple layers of features from tiny images. Technical Report, University of Toronto (2009)
10. Zagoruyko, S., Komodakis, N.: Wide residual networks. In: British Machine Vision Conference (2016)
11. Sutskever, I., et al.: On the importance of initialization and momentum in deep learning. In: International Conference on Machine Learning. PMLR (2013)
12. Polyak, B.T.: Some methods of speeding up the convergence of iteration methods. USSR Comput. Math. Math. Phys. **4**(5), 1–17 (1964)
13. Abuduweili, A., et al.: Adaptive consistency regularization for semi-supervised transfer learning. In: Proceedings of the IEEE/CVF Conference on Computer Vision and Pattern Recognition (2021)
14. Fan, Y., Kukleva, A., Dai, D., et al.: Revisiting consistency regularization for semi-supervised learning. Int. J. Comput. Vision **131**(3), 626–643 (2023)
15. Chen, B., et al.: Debiased self-training for semi-supervised learning. Adv. Neural Inf. Process. Syst. (2022)
16. Long, J., et al.: A novel self-training semi-supervised deep learning approach for machinery fault diagnosis. Int. J. Prod. Res. **61**, 1–14 (2022)
17. Zhang, Y., et al.: Deep long-tailed learning: a survey. IEEE Trans. Pattern Anal. Mach. Intell. (2023)
18. Zhang, C., et al.: An empirical study on the joint impact of feature selection and data resampling on imbalance classification. Appl. Intell. **53**(5), 5449–5461 (2023)
19. Wang, W., et al.: Imbalanced adversarial training with reweighting. In: 2022 IEEE International Conference on Data Mining (ICDM). IEEE (2022)
20. Li, J., Liu, Y., Li, Q.: Generative adversarial network and transfer-learning-based fault detection for rotating machinery with imbalanced data condition. Meas. Sci. Technol. **33**(4), 045103 (2022)
21. Shi, Y., et al.: Improving imbalanced learning by pre-finetuning with data augmentation. In: Fourth International Workshop on Learning with Imbalanced Domains: Theory and Applications. PMLR (2022)
22. Bonner, S., et al.: Implications of topological imbalance for representation learning on biomedical knowledge graphs. Brief. Bioinf. **23**(5), bbac279 (2022)
23. Gouabou, A.C.F., et al.: Rethinking decoupled training with bag of tricks for long-tailed recognition. In: 2022 International Conference on Digital Image Computing: Techniques and Applications (DICTA). IEEE (2022)

Online Learning Behavior Analysis and Achievement Prediction with Explainable Machine Learning

Haowei Peng, Xiaomei Yu$^{(\boxtimes)}$, Xiaotong Jiao, Qiang Yin, and Lixiang Zhao

Shandong Normal University, Jinan, China
yxm0708@126.com

Abstract. With the development of computer technology, "Internet + education" has become a hot topic, which has promoted the development of online education. The impact of the epidemic has brought new challenges to online education, and the effect of online teaching has sparked extensive research. Why online learning has not received satisfying learning effect? And how to improve the quality of online education have become the focus of research. To address these challenges in online learning, this paper studies the important factors affect the students' learning achievements based on the explicit and implicit features extracted from the online learning behavior data. And a wealth of machine learning models are explored to predict the learner's achievement, including the original linear models, the ensemble learning models and the advanced deep learning models. With extensive experimental results achieved, we gain reasonable explanation on the influence factors hander the learning achievement, and predict the learning achievements with the explainable machine learning models. Finally, some suggestions are put forward to improve students' learning behaviors, aiming to promote the quality of online learning in the long run.

Keywords: Machine learning · Achievement prediction · Learning achievement · Learning behavior data

1 Introduction

With the rapid development of "Internet +" technology and 5G, online education with Internet + technology has become a popular trend in modern education [1]. Artificial intelligence is widely used in medicine, education, the military, etc [2]. Compared with the traditional learning mode, online learning are not limited in the classroom capacity, time or space, which promotes the concept of "learning anytime, anywhere" to come true. At the beginning of 2020, the sudden outbreak of COVID-19 has brought great impact to all walks of life. For online education, this is both an opportunity and a challenge. In accordance with the requirements of "class suspension without school suspension", online learning boosts the first large-scale learning and teaching activity using Internet technology in the history

J. Vaidya et al. (Eds.): AIS&P 2023, LNCS 14510, pp. 22–37, 2024.
https://doi.org/10.1007/978-981-99-9788-6_3

of higher education in China [3]. According to the survey, during the epidemic, a total of 423 million users in China carried out teaching and learning on the Internet, and various MOOC platforms have been unprecedentedly promoted and popularized [4].

Why online learning has not received satisfying effect? And how to improve the quality of online education? To address these challenges, we have collected a large amount of learning behavior data from online learners and extracted various explicit and implicit features for learning behavior analysis. After data preprocessing, a wealth of machine learning models are explored in predicting the learner's achievement, including the original linear models, the ensemble learning models and the advanced deep learning models. With extensive experimental results achieved, we endeavor to settle the following three problems with reasonable explanation:

- Based on the online learning behavior data obtained, which machine learning model is effective to predict the learner's achievement, in details, pass or fail the course in the long run?
- With the explicit and implicit features extracted, which are the Top-K important factors affect the learners' achievement in online course?
- How degree does each of the Top-K important factors influence the learners' achievement?

The rest of the paper are as follows. In Sect. 2, the related work on machine learning-based education are overviewed, including the achievements with explainable machine learning models and unexplainable deep learning models. And then, our work based on the investigation is introduced in Sect. 3. In Sect. 4, a series of experiments are conducted and the experimental results are analyzed. Still then, the educational application with experimental results is illuminated in Sect. 5. Finally, the study is concluded and the future directions are pointed out in Sect. 6.

2   Related Work

This study adopts three types of machine learning models to explore the learners' achievement, including Support Vector Machine (SVM), Logistic Regression (LR), Random Forest (RF), AdaBoost and Long Short-Term Memory (LSTM). By comparing various models, the best model is selected to predict students' grades and provide guidance for teaching activities.

2.1   Explainable Single Models

Logistic Regression. Logistic Regression is a linear classifier, which is generally used to solve classification problem. As an explainable machine learning model, it is mainly used to explore happens of some events in form of probability. Specifically, be confronted with a regression or classification problem, one has to establish the cost function, and then use the optimization method to iteratively discover the optimal parameters of the model, and still then to test and verify the solution model before applied to real world applications.

According to the characteristics of Logistic Regression model, which is good at solving classification problems, researchers use it to explore the possible causes of some events. Zong et al. summarized the research process of Logistic Regression in the field of online learning, analyzed the learning data of MOOCS with Logistic Regression, explored the impact of learners' online learning behavior on learning achievement, and constructed the Logistic Regression prediction model [5]. Dai et al. used the Logistic Regression method to comprehensively analyze whether the graduation design of continuing education students can meet the requirements of graduation achievement [6].

Support Vector Machine. Support Vector Machine algorithm is a supervised machine learning algorithm. Support Vector Machine is a generalized linear classifier for binary classification of data [7]. SVM was the first classifier developed from the generalized portrait algorithm in pattern recognition, which uses the inner product kernel function to transform the multi-dimensional nonlinear problem into two-dimensional nonlinear problem. In calculation, the SVM only depends on a small number of key samples, which avoids the interference of spatial dimension. Without the calculation difficulty encountered in some models due to the number of samples, the SVM has largely reduced calculation, and results in its good robustness [8].

However, the kernel function of SVM is sometimes difficult to determine in dealing with nonlinear problems. In some extent, the SVM can be regarded as a network with a hidden layer, and the number of neurons in the hidden layer is equal to the number of support vectors. Therefore, the SVM does not have the interpretability inherently as the linear models and Decision Tree do.

Educational research based on SVM algorithm has attracted extensive attention in recent years. L. Ma used SVM algorithm to predict the failure of students' current courses, so as to improve students' learning quality and efficiency [9]. Zhang et al. proposed a prediction model of postgraduate entrance examination in colleges and universities based on SVM to predict the results or achievements of postgraduate entrance examination, so as to help students with thier decision-making [10].

2.2 Explainable Ensemble Models

An ensemble model performs the learning task by combining multiple basic machine learning models, and achieves its good performance with the effectiveness of all the single models jointly. Given a data set, it trains several individual learners respectively to form an integrated one with a certain set of strategies. Combining the strengths of each learner, an ensemble model obtains better generalization performance than a single basic learner [11].

Random Forest. Random Forest model is an integrated learner commonly used in communities of academia and industry. In Random Forest model, multiple subtrees are constructed to train the samples random selected, and realizes the summary of prediction results of multiple decision trees by introducing voting mechanism. The output category of a Random Forest classifier is determined by the number of output categories of individual decision trees, which combines Breimans' idea of "Guided aggregation" with Ho's "Random subspace method" [12]. The Random Forest converges to smaller generalization error with fast generalization speed and high training efficiency. Moreover, the RF classifier is strong to resist the interference of missing data, and wins with high accuracy.

Integrating multiple base learners of decision trees with inherently interpretability, the Random Forest model has trustable analysis effect on multidimensional features. Generally speaking, the interpretability of Random Forest model and its explanation on the importance of features can be achieved by various tools, such as PFI (Permutation Feature Importance) methods and LIME (Local Interpretable Model-agnostic Explanations) models.

Many experiments have carried out in the field of education using Random Forest algorithm and its interpretability, and achieved good results. Niu et al. used Random Forest algorithm to find out the main characteristics and some causes that affecting employment, and provided better support strategies for the employment of private college graduates [13]. Lv et al. used the basic data of students' learning trajectory data and the Random Forest algorithm to analyze the causes of students' academic early warning, and give the relevant countermeasures for the construction of academic early warning mechanism [14].

AdaBoost. The AdaBoost algorithm is constructed on base classifiers with various methods, and it provides a framework for model combination and performance promotion. Considering that the construction of each base classifier is simple and its performance is weak, it is not worth the feature screening with each single one, since the performance of an AdaBoost classifier can be promoted by adding a new classifier without fall into the dilemma of over-fitting.

The AdaBoost algorithm is suitable for binary classification or multi original classification scenarios, and is capable of feature selection. As each base classifier is simple explainable machine learning models, the interpretability of AdaBoost is achieved with interpretable machine learning library such as ELI5 and SHAP (SHapley Additive exPlanation).

AdaBoost algorithm is often used in classification and prediction problems. Zhao analyzed the shortcomings of the existing system and used AdaBoost algorithm to train students' academic level prediction model [15]. Liu compared the results of different models, and finally used AdaBoost to predict student performance, and applied the performance prediction to the specific student management system [16].

2.3 Unexplainable Deep Learning Models

As a currently prevailing machine learning model, the deep learning is capable of exploring internal law and representation level from a large amount of samples. With big data available and multi-dimension features extracted, deep learning models make remarkable achievements in field of image processing, natural language processing (NLP) and so on. However, constructed on deep neural networks with thousands of neurons, a general deep learning model consists of multi hidden layers with complex structure. The interpretability of deep learning models is still unsolved which deserves the joint efforts of scientists.

As a widely used deep learning model, the Long Short-Term Memory network is structured on time recursion, which is skilled in dealing with time series data. It is specially designed to solve the long-term dependence limitation in general RNN (Recurrent Neural Network). With the special memory units arranged in an RNN structure, the LSTM is verified to be capable of storing sequence information for a relate long time [17]. The LSTM network has made outstanding contributions in the fields of natural language processing, speech recognition, image translation, timing analysis and machine translation. However deep learning models are prone to severe overfitting in training and poor classification accuracies for some categories [18].

LSTM algorithm has also been widely used in the field of education in recent years. Aiming at the timing of students' historical achievements and the forgetting characteristics of learning process, Cao et al. introduced LSTM network to model the state of students' knowledge structure [19]. Yin et sl proposed an emotion classification fusion model based on LSTM and GCNN [20]. Cheng et al. proposed an early warning model based on LSTM optimized neural network, aiming to solve the low prediction accuracy and poor generalization of the existing system [21].

2.4 Comparison on the Above Five Models

The deep learning methods are suitable for dealing with big data. It is proved that deep learning is slightly inferior to traditional machine learning in performance when the data scale is the same and the data scale is small. As shown in Fig. 1, as the data scale reaches a certain degree, the traditional machine learning model encounters a performance bottleneck, while the deep learning demonstrates its overwhelming advantages in performance gradually. Therefore, when the amount of data available is relatively small, the traditional machine learning methods would be the dominant ones to reveal remarkable performance.

In general, the traditional machine learning models have few parameters and limited feature space. As there is a large amount of data, it is difficult to learn the complex high-dimensional relationships with some machine learning methods, and they fall in performance bottlenecks. On contrast, the deep learning models have many parameters, strong representation ability and easy over fitting when confronted with a data set of insufficient samples. Therefore, the performance of the deep learning model is slightly inferior to the traditional machine learning

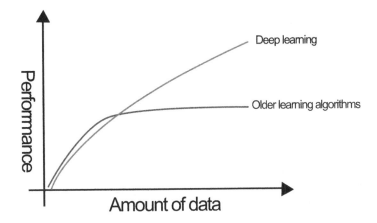

Fig. 1. Relationship between performance and the amount of data.

model when dealing with the data in general volume. Moreover, as to the execution time that a model spent in training, a deep learning model would spend much more time than a machine learning one. Since the deep learning model contains more parameters, it takes longer to train them than a machine learning one.

3   Method

To address the three research questions, we explore the effective machine learning models to predict the learner's achievement, based on the online learning behavior data obtained. With the best model discovered, we utilize explainable tools to reveal the Top-K important factors that affect learning achievement in online course. And then, focusing on each important factor in the Top-K, we study the degree of its influence on the learners' achievement with ablation experiment. The construction process of the models is shown in Fig. 2.

3.1   Learning Achievement Prediction

In the process of educational informatization, establishing a performance prediction model helps teachers and students adjust teaching content and progress in a timely manner. For students, performance warning can help them adjust learning strategies and improve learning outcomes in a timely manner. For teachers, grade prediction is convenient for them to pay attention to students' grades, timely intervene in the early warning of marginalized students' grades, and improve teaching effectiveness. In summary, the performance prediction model helps to carry out teaching activities, thereby reducing the number of failed students and improving overall academic performance.

Firstly, the collected data are preprocessed. With missing data discarded and effective features extracted, we construct the dataset for downstream tasks.

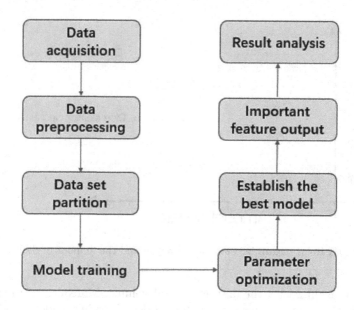

Fig. 2. Achievement prediction model.

Secondly, as the learning behavior data set is divided into training set and test set, each model is trained with the training set including a large amount of samples. After the process of training and test on each model, we record the optimal parameters and calculate the experimental result. Finally, comparing all models with each other, we obtain the most effective one on the metric of predictive performance.

3.2   Top-K Influence Factors Discovery

The experimental results on learning achievement prediction reveal the possible learning effect with the current learning behaviors, while there is no reasonable explanation on the students' temporary learning failure. Therefore, nobody can tell the trustable reasons lead to his current learning failure. To address the second question, we utilize explainable tools to explore the Top-K features that determine the prediction performance of the most effective explainable machine learning model. Truly speaking, these Top-K features are the most important factors that have obvious effect on in learner's achievement.

With the Top-K important factors that influent the learning achievement discovered, a student can check his learning behavior and find out the possible reasons that hinder his learning achievement. In this way, the direction on improving learning effect is available that deserve to have a try.

3.3 Importance of Each Factor

After find out the Top-K important factors that affect the learner's achievement, the further study is necessary to explain the influence of each factor for the learning achievement. Therefore, the ablation experiment is introduced to illustrate the degree of each Top-K factor response for the performance degradation in the prediction experiments. Specifically, we discard each Top-K feature in data respectively, and conduct the prediction successively. Comparing the extent of performance degradation in probability, the influence of each Top-K factor is gained which lights up the obvious relationship between each learning behavior and the learner's achievement.

The output of ablation experiments verifies the effectiveness and influence of the Top-K important factors in form of probability, and provides the teachers' teaching and students' learning with guide in their learning activities. With the relationship of learning behavior and achievement released and influence of each factor discovered, the student may increase his confidence on the explainable and measureable improvement that contribute to his learning achievement. For the teachers, it is necessary to help the students to perform their phased and hierarchical improvement in learning.

4 Experiment

4.1 Learning Achievement Prediction

Data Set. With the popularization of online education and the rapid development of educational information systems, a large amount of educational and teaching data has been presented and saved [22]. The experimental data set came from an online teaching platform used by an university. A total of 2234 students' learning behavior data are collected. The data were generated as the students were learning a course on the online learning platform, which represent the various characteristics of students' learning behavior from the behavior logs. Specifically, the data describe the students' effective learning behaviors mainly include the following aspects:

Watch Course Videos. It is the main way for the students to learn the online course. By playing video resources, students intuitively follow the teachers to learn course content.

Submit Assignments and Quizzes. After the course content is explained, the teachers assign some tasks for students to learn to consolidate their learning achievements. During the course, teachers would organize some tests, so as to grasp the students' learning effect. At the same time, the students can check and make up for deficiencies according to the test results.

Post Topic and Reply in the Discussion Area. After learning the videos, the students may still have some problems unsettled. At this time, they could go to the discussion area corresponding to the course and search for answers. Otherwise, the question can be post for help. In summary, each effective learning behavior in the course discussion was recorded for learning achievement analysis.

Other Learning Behaviors. When the students registered the course, the teachers would provide the students with other forms of learning resources including course videos, such as the courseware, bibliography and so on, which were used as a supplement to video learning and are collected as the important learning behavior data in the experiments.

Data Preprocessing. Due to the large amount data of complex types in the data set, there are problems such as missing values, outliers and inconsistent data format. Therefore, it is necessary to preprocess the data before model training, so as to simplify data and improves efficiency [23], and eventually promote the accuracy of the model in training and prediction. The data preprocess in our experiments includes the following steps.

Missing Value Processing. The default student scores are meaningless to the construction of the model for learning achievement prediction, so the records of these students are deleted directly.

Independent Variables. Aiming to build a model for learning achievement prediction, we delete the variables irrelevant to the tasks such as "Class", "Grade", "StudentID" and so on.

Model Training. Before model training, the data are divided into the training set and the test set, and then one feature is set as the target feature and the rest as the attribute ones. Continually, the classification trains the data on the training set, and tests the effect of the prediction model with the test set data. Finally, we evaluate the prediction performance of each model according to the model evaluation index.

In our experiments, the original data set is divided into training set and test set according to the ratio of seven vs. three. The feature of "Score" is taken as the target eigenvalue, while the remaining features are regarded as attribute eigenvalues excluding the features of "Class", "Grade" and "StudentID".

Parameter Optimization. The method of grid search is adopted to optimize the parameters in the model automatically. For example, with 670 samples available for model tuning, the optimal parameters in random forest model are output. After learning achievement prediction with the model, the confusion matrix is achieved based on the experimental results obtained, as shown in Table 1. It demonstrates that the RF model has achieved effective performance.

Table 1. Confusion matrix

	Pred-pass	Pred-fail
Pass	574	10
Fail	19	67

Comparison with Other Models

Evaluating Indicator. We use several evaluation metrics to evaluate the performance of the models, including Accuracy, Recall and F1-score.

Accuracy refers to the ratio of the number of samples correctly classified by the classifier to the total number of samples in a given test data set. That is, the accuracy on the test data set when the loss function is 0–1 loss. The Accuracy is calculated as follows:

$$Accuracy = \frac{TP + TN}{TP + TN + FP + FN} \tag{1}$$

Recall rate: in the prediction results, the Recall rate describes the situation in proportion that how many positive samples are correctly predicted in the original data set, which is represented as follows:

$$Recall = \frac{TP}{(TP + FN)} \tag{2}$$

Precision rate: in the prediction results, the Precision rate describes the situation in proportion that how many really positive samples are correctly predicted in the positive samples of the original data set, which is represented as follows:

$$Precision = \frac{TP}{(TP + FP)} \tag{3}$$

F1-score: As a harmonic average of the Precision and the Recall, the F1-score is calculated with (4) as follows:

$$F1 = \frac{2 \times P \times R}{P + R} \tag{4}$$

Comparison results

We conduct the experiment for learning achievement prediction with five machine learning models of three types, including the Logistic Regression, AdaBoost, Support Vector Machine, Random Forest, and LSTM model. Each model is applied for prediction on the same data set respectively. And the metrics of Accuracy, Recall and F1-score are recorded and compared with each other.

As shown in Table 2, some observations are achieved as follows: the explainable ensemble models win advantageous performance in learning achievement prediction. The Accuracy rate of Random Forest model reaches 96%, which is

Table 2. Model comparison results

Model	Accuracy	Recall	F1-score
LSTM	0.904	0.500	0.474
LR	0.912	0.674	0.663
Adaboost	0.937	0.744	0.753
SVM	0.910	0.662	0.655
RF	*0.958*	*0.779*	*0.822*

much higher than the other three models of LSTM, LR and SVM. Moreover, the Recall rate and F1-score in RF model are also higher than that of the other three models.

The deep learning model of LSTM achieves effective performance in learning achievement prediction. However, it does not win with remarkable performance in dealing with the time series data in the prediction tasks. Considering that the unexplainable deep learning models show their strengths at learning based on a large amount of data, the LSTM model is limited with insufficient data and does not give full play to advantages in dealing with the thousands of samples in a course.

As explainable single models of LR and SVM, both of them show the high Accuracy and acceptable Recall in the experiments. We explore the potential causes and discover that some positive samples of are incorrectly labeled as negative samples under the strict requirements on achievement of "Score", with the purpose to inspire learning and to urge the most students along. However, the incorrectly labeled negative samples weaken the ability of the model to learn positive examples, so as to predict some of the positive samples as negative ones. Considering that it may encourage the "probably failed" students to improve learning behaviors and promote learning efficiency as early as possible, we regard it acceptable in Recall.

With comprehensive consideration on the explainability and performance of the machine learning models in learners' achievement prediction, we use the Random Forest model to exert prediction task and Top-K factors affecting learning effect in the second downstream tasks.

4.2 Top-K Factors Affecting Learning Effect

The interpretability of the model is conducive to obtain more useful information, and explain the causality in the data, and eventually win the general trust in the applications [24]. In our research, it is necessary not only to predict the performance, but also to identify the important factors determining the learners' achievement.

The Random Forest model is not only widely used in prediction, but also often applied to feature selection. After the establishment of the learning achievement prediction model, we exert the Random Forest algorithm to further explore the learning behaviors that have a great impact on the learning achievement, so as to help teachers and students improve learning activities purposefully.

After the data fitting, the measurement of feature importance rank is listed according to the attribute data, which benefits from the parameter of importance in the Random Forest model. This parameter returns objects in form of numpy array, corresponding to the features' importance in the training data set with the Random Forest model. The values of `feature importances` returned are of float type less than 1. In the feature importance array, the attribute column with the higher value is more important for the performance in prediction.

Visualize the importance features of the model and the results is shown in Fig. 3, in which the abscissa represents the features and the ordinate denotes the weights. As there are so many features in the data set and some features is of little importance to the effectiveness in prediction, we set the `threshold` to 0.05 and output the features with importance weight greater than 0.05, six eligible features were selected. Some conclusions can be drawn with the experimental results. As the answer to the second question in the paper, the factors of feature 4, feature 26, feature 6, feature 19, feature 15 and feature 0 have a great impact on the students' achievements. Specifically, the Top-6 factors of students' learning behaviors affecting the learning achievement are sorted from high to low, they are the Total browsing time, Task submission times, Discussion frequency, Stage test score, Browsing count and Task score. Among them, the factor of Total browsing time has the greatest impact on students' academic performance, and it reaches 0.2359.

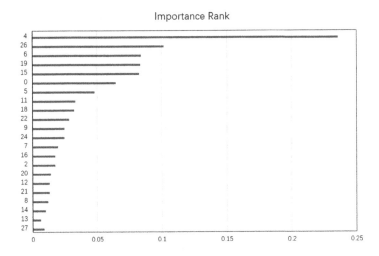

Fig. 3. Feature importance rank.

4.3 Verification of Important Features

In order to explore the influence of each factor on the learning effect, and further to make guiding suggestions to teachers and students, we conduct ablation experiments on the RF model. "Ablation study" is applied to describe the process of removing some features before model learning, so as to better understand the behavior of the model and the role of the features removed. We process the original data set by removing the top six features in the feature importance rank, and compare the performance of the model on original data set with the same one on feature removed data set (Fig. 4). As shown in Table 3, the accuracy of the RF model is reduced since we remove the features with high importance weight. Taking the Total browsing time as an example, we removed the data of this feature from the data set. At this time, the accuracy of the random forest model decreased significantly compared to the original data, indicating that this feature has a significant impact on the prediction results. Furthermore, the higher the weight of the feature importance is, the more obvious the performance of

Table 3. Ablation experimental results

	Accuracy	Recall	F1-score
Original model	0.958	0.779	0.822
Total browsing time	0.949	0.709	0.782
Task submission count	0.950	0.727	0.789
Discussion frequency	0.950	0.735	0.802
Stage test score	0.952	0.756	0.805
Browsing count	0.953	0.762	0.809
Task score	0.955	0.773	0.813

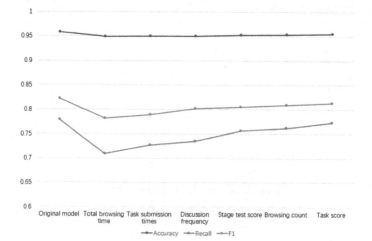

Fig. 4. Ablation experimental results.

RF drops. The results of ablation experiments show that the feature importance rank is effective and explainable, which indicates the direction for teachers and students on how to improve the teaching and learning activities in the course.

5 Educational Application

Against the backdrop of the rapid development of information technology, it is clear that applying information technology to the field of education and improving teaching has become a trend [25]. Based on the analysis of experimental results, we can conclude that the Total browsing time, Task submission count, Discussion frequency and other factors in online course learning have a great impact on the learners' achievement. It is necessary for the teachers and students to pay attention to the above-mentioned aspects in their daily learning, so as to improve the learners' achievement.

The total browsing time of the course is directly proportional to the student's academic performance, and the longer the browsing time, the better the score. Because when designing a course, in addition to focusing on the course content, it is also necessary to combine the learning characteristics of students to attract their attention. On the screen, online learning refers to students learning anytime and anywhere by watching videos, which tests their willpower more than offline in the classroom. Therefore, when designing videos, more animation styles can be added to make the screen more vivid to maintain students' attention.

For teachers, adopting task driven teaching in classroom activities and post class discussions is acceptable, as it can promote students' active participation in learning. Before class, the teacher can propose a topic related to the course content to arouse students' interest. Students discuss the knowledge they have learned, guide themselves into the course materials, and then engage in collaborative learning. This not only benefits classroom progress, but also helps students understand and think about the knowledge they have learned.

In order to increase the number of task submissions for students, teachers can change the current course evaluation method and include the number of task submissions in the final academic performance assessment. For example, they can weight their daily performance and final exam scores in a ratio of 3:7 to obtain the final score. This way, students can actively participate in their daily course tasks and gradually complete the course goals.

6 Summary

The learners' achievement is the direct embodiment of learning effect. By constructing explainable machine learning models to predict the students' achievement, this paper explores the performance of five machine learning models in prediction model and discover the Top-K important factors which influence the students' learning achievement. With explainable machine learning tools and ablation experiments, we point out the influence of each Top-K factor on the students' achievement, and further put forward some suggestion on teaching and learning activities in course learning. However, with the limited amount of data available, the LSTM model did not demonstrate its good performance in prediction, and the explainable machine learning model proposed in this paper needs further research on big data with amply learning behavior records. These are the future research directions deserve further studied.

Acknowledgments. This work is supported by Shandong Provincial Project of Graduate Education Quality Improvement (No. SDYJG21104, No. SDYJG19171), the Key R&D Program of Shandong Province, China (NO. 2021SFGC0104, NO. 2021CXGC010506), the Natural Science Foundation of Shandong Province, China (No. ZR2020LZH008, ZR2021MF118, ZR2022LZH003), the National Natural Science Foundation of China under Grant (NO. 62101311, No. 62072290), the Postgraduate Quality Education and Teaching Resources Project of Shandong Province (SDYKC2022053, SDYAL2022060), the Shandong Normal University Research Project of Education and Teaching(No. 2019XM48), and Industry-University Cooperation and Education Project of Ministry of Education (No. 2206026 95231855).

References

1. Yu, X.: Software and Educational Information Service. Shandong People's Publishing House, Jinan (2022)
2. Wang, T., Zheng, X., Zhang, L., Cui, Z., Xu, C.: A graph-based interpretability method for deep neural networks. Neurocomputing **555**, 126651 (2023)
3. Wu, D., Li, W.: Stage characteristics of large-scale online teaching in Chinese universities: empirical research based on group investigation of students, faculty and academic staff. J. East China Normal Univ. (Educ. Sci.) **38**(07), 1–30 (2020)
4. Xiao, E., Xia, F.: Research on the impact of Covid-19 on the development of online education in colleges and universities - taking Hubei Engineering University as an example. Stat. Consult. **5**, 4 (2021)
5. Zong, Y., Sun, H., Zhang, H., Zheng, Q., Chen, L.: Logistic regression analysis of learning behavior and learning effect of MOOCS. Educ. Sci. Abs. **35**(4), 2 (2016)
6. Dai, J., Fang, L., Li, J., Jin, L., Song, L., Chen, X.: Identification of test score associated factors based on logistic regression. Exam. Res. **04**, 70–74 (2019)
7. Zhou, Z.: Machine Learning. Tsinghua University Publishing House Co., Ltd., Beijing (2016)
8. Cheng, H.: Research on academic early-warning system based on improved LSTM algorithm. Guilin Univ. Electron. Technol. (2021). https://doi.org/10.27049/d.cnki.ggldc.2021.000113
9. Ma, L.: Research and application of student academic early warning based on convolutional neural network and SVM (2021). https://doi.org/10.27232/d.cnki.gnchu.2021.002175
10. Zhang, K., Yan, L., Liu, C.H., Du, Y.: Prediction model of postgraduate entrance examination based on SVM. J. Henan Univ. Urban Constr. **30**(6), 7 (2021)
11. Shi, C.H., Wang, X., Hu, L.: Introduction to Data Science. Tsinghua University Publishing House Co., Ltd., Beijing (2021)
12. Lipton, C.Z.: The mythos of model interpretability. Commun. ACM **61**(10) (2018)
13. Niu, D., Liu, J.: Research on employment early warning mechanism based on random forest algorithm. Mod. Inf. Technol. **5**(22), 3 (2021)

14. Lv, L., Xia, Z.H.: Cause analysis and countermeasure research of college academic warning based on random forest algorithm. J. Nanchang Inst. Technol. **39**(6), 81–86 (2020)
15. Zhao, X.: Research and application of students' academic level prediction based on adaboost. Northeastern University, MA thesis (2016)
16. Liu, X.: Research and Application of Performance Prediction Model Based on Student Behavior. University of Electronic Science and Technology of China, Chengdu (2017)
17. Yang, L., Wang, Y.: Survey for various cross-validation estimators of generalization error. Appl. Res. Comput. **32**, 1287–1290, 1297 (2015)
18. Chen, X., Zheng, X., Sun, K., Liu, W., Zhang, Y.: Self-supervised vision transformer-based few-shot learning for facial expression recognition. Inf. Sci. **634**, 206–226 (2023)
19. Cao, H., Xie, J.: Research on learning achievement prediction and its influencing factors based on LSTM. J. Beijing Univ. Posts Telecommun. (Social Sci. Ed.) **22**(6), 11 (2020)
20. Yin, Y., Zheng, X., Hu, B., Zhang, Y., Cui, X.: EEG emotion recognition using fusion model of graph convolutional neural networks and LSTM. Appl. Soft Comput. **100**, 106954 (2021)
21. Cheng, H., OuYang, N., Lin, L.: Application of an LSTM optimization algorithm in early warning of college students' academic performance. Mod. Electron. Techn. **45**(10), 142–147 (2022)
22. Chen, Q., Yu, X., Liu, N., Yuan, X., Wang, Z.: Personalized course recommendation based on eye-tracking technology and deep learning. In: 2020 IEEE 7th International Conference on Data Science and Advanced Analytics, Sydney, NSW, Australia, pp. 692–968 (2020)
23. Romero, C., Ventura, S.: Data mining in education. Wiley Interdisc. Rev. Data Min. Knowl. Disc. **3**(1), 12–27 (2013)
24. Kabakchieva, D.: Student performance prediction by using data mining classification algorithms. Int. J. Comput. Sci. Manag. Stud. **1**(4), 686–690 (2012)
25. Jiao, X., Yu, X., Peng, H., Zhang, X.: A smart learning assistant to promote learning outcomes in a programming course. Int. J. Softw. Sci. Comput. Intell. **14**(1), 1–23 (2022)

A Privacy-Preserving Face Recognition Scheme Combining Homomorphic Encryption and Parallel Computing

Gong Wang[1], Xianghan Zheng[1], Lingjing Zeng[2], and Weipeng Xie[1(✉)]

[1] Fuzhou University, Fuzhou, China
2431455652@qq.com
[2] Fujian Chuanzheng Communicatians College, Fuzhou, China

Abstract. Face recognition technology is widely used in various fields, such as law enforcement, payment systems, transportation, and access control. Traditional face authentication systems typically establish a facial feature template database for identity verification. However, this approach poses various security risks, such as the risk of plaintext feature data stored in cloud databases being leaked or stolen. To address these issues, in recent years, a face recognition technology based on homomorphic encryption has gained attention. Based on homomorphic encryption, face recognition can encrypt facial feature values and achieve feature matching without exposing the feature information. However, due to the encryption, face recognition in the ciphertext domain often requires considerable time. In this paper, we introduce the big data stream processing engine Flink to achieve parallel computation of face recognition in the ciphertext domain based on homomorphic encryption. We analyze the security, accuracy, and acceleration of this approach. Ultimately, we verify that this approach achieves recognition accuracy close to plaintext and significant efficiency improvement.

Keywords: Homomorphic Encryption · Face recognition · Flink · Privacy Protection · Data Flow

1 Introduction

As the era of the Internet of Things approaches, various smart sensing devices are widely used, and face recognition systems are widely applied in areas such as law enforcement, payment, transportation, access control, and more. Compared to traditional identity authentication mechanisms, face recognition features the advantages of being less prone to forgetting or damage and offers convenient usage, effectively enhancing the efficiency of authentication. One common attack on face recognition is the use of facial feature template libraries. For instance, attackers can reconstruct the original face image to successfully bypass the authentication or infer personal attributes like age and gender.

Since data involves sensitive information, uploading data to the cloud poses risks of leakage. An effective solution is to encrypt the data before uploading it,

© The Author(s), under exclusive license to Springer Nature Singapore Pte Ltd. 2024
J. Vaidya et al. (Eds.): AIS&P 2023, LNCS 14510, pp. 38–52, 2024.
https://doi.org/10.1007/978-981-99-9788-6_4

which can prevent data leaks, but it hinders feature matching in the ciphertext domain. General encryption algorithms can only perform encryption, and it has become a hot topic in academia to find a solution for performing computations on encrypted data.

Enabling computation on encrypted data has been a challenging issue that has perplexed many scholars until the advent of Fully Homomorphic Encryption (FHE) schemes. FHE can perfectly solve this problem as it allows computations on encrypted data. Subsequently, many scholars, inspired by his work, have gradually developed various FHE schemes, enriching the field of encrypted computation and providing theoretical support for achieving face recognition in the ciphertext domain.

In comparison to the existing homomorphic encryption face recognition algorithms, which are extremely time-consuming, this paper focuses on the challenges faced by face recognition in the ciphertext space, such as low efficiency and massive computational overhead. To address these challenges, the paper proposes a data stream computing framework for face privacy protection. It encrypts face data using the CKKS homomorphic encryption scheme and processes the data as a data stream using Flink, a stream processing framework. Flink features real-time processing, multi-threading, and stateful data processing, dividing the stream data processing into two Map-Reduce stages. The proposed framework aims to protect the facial ciphertext information stored in the database, enabling approximate homomorphic encryption for matching calculations while utilizing Flink for stream processing to improve the efficiency of facial ciphertext computation. This approach ensures both the security of facial data and high efficiency and practicality in the process.

2   Related Work

2.1   Homomorphic Encryption

Homomorphic encryption is one of the important techniques for privacy protection. The idea behind homomorphic encryption is to perform encryption first and then perform encrypted calculations based on the ciphertext data. The result of the encrypted computation, when decrypted, will match the result of the same computation on the original unencrypted data.

In 2017, Cheon et al. proposed the CKKS (Cheon-Kim-Kim-Song) homomorphic encryption scheme [1]. It is an approximate homomorphic encryption scheme that supports floating-point number calculations and has high efficiency. Currently, it is one of the most widely used homomorphic encryption schemes [2].

As the theory of homomorphic encryption continues to evolve, this technology has been widely applied in various fields [2–6], such as machine learning [7,8], secure multi-party computation [9–11], federated learning [12,13], and cloud computing [14–16]. However, the high computational overhead and low efficiency of homomorphic encryption privacy protection schemes have been major bottlenecks restricting the development of homomorphic encryption technology. In recent years, many experts and scholars have been dedicated to improving the efficiency of homomorphic encryption technology [17–21]. Parallel computing techniques have been widely used to enhance the execution efficiency of homomorphic encryption schemes, such as ciphertext packing techniques [22], batch processing, and single instruction multiple data (SIMD) techniques, among others.

2.2 Flink Stream Processing Engine

Apache Flink is a distributed processing engine primarily designed to handle streaming data and supports stateful computations. In the Flink framework, all data is treated as data streams, where batch data represents bounded data streams and real-time data means unbounded data streams. As a result, Flink is a unified big data processing engine that can perform stream and batch processing operations. One of the critical features of Flink is that it is event-driven. It can receive data from multiple data sources as data streams and triggers corresponding data operations only when data is received. No procedures are performed when data is not available.

2.3 FaceRecognition

FaceRecognition is an open-source face recognition library based on the Python programming language. It utilizes a simple API to perform face detection and recognition. The library offers a range of powerful functionalities, including face alignment, feature extraction, and more, making it suitable for various applications such as face recognition and face verification.

3   Methodology

3.1   Methodology Overview

In this paper, the features extracted from facial images are first encrypted and stored in a cloud-based database. During face identity verification, the features of the current face are extracted and encrypted. The feature matching phase utilizes Flink for data stream processing, which consists of two sub-stages. The Map operator calculates the differences between encrypted data streams for each sample and then squares them. The Reduce operator aggregates the results of all data streams and computes the sum. Finally, the encrypted comparison score between the two is returned.

This scheme aims to protect the private information of users stored in the template database by performing approximate homomorphic encryption for matching calculations. Simultaneously, it utilizes Flink for data stream processing to improve computational efficiency. The approach ensures both data security and efficiency, providing a practical and effective solution.

System Model: In privacy-preserving machine learning algorithms, one of the most common computations is the matrix multiplication between two matrices. Matrices can be viewed as special types of vectors, which allows us to transform the matrix multiplication into the form of a vector inner product. The vector inner product involves the multiplication of corresponding elements in two vectors followed by their summation. The element-wise multiplication process between vectors is independent of each other, making it amenable to parallelization to improve efficiency. Finally, the results of parallel computations are aggregated and summed up. The framework can be seen in Fig. 1.

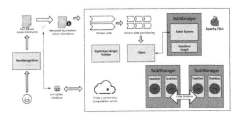

Fig. 1. Frame Work

3.2   Feature Encryption

To prevent malicious attacks on the template database, which could lead to obtaining real facial images and their corresponding feature attributes, it is necessary to protect the stored feature templates in the cloud. Therefore, the proposed scheme suggests encrypting the successfully extracted feature templates using approximate homomorphic encryption and then sending the encrypted feature templates to the cloud. The key generation and encryption process for approximate homomorphic encryption are as follows:

Key Generation: Use the SEAL library to get the locally transmitted parms build parameter container. The CKKS framework is then generated using the parameter params: contextSEALContext context(params). Obtain the key list through local transmission. After the key is generated, the decrypted private key is saved by the local client by calling the HE.save_key key saving function in the homomorphic encryption module.

The public key, relinearized key, and rotating key need to be sent to the cloud for subsequent homomorphic calculation. Relinearized key is mainly used to reduce noise. Euclidean distance calculation requires the use of rotation keys. The algorithm flow can be seen in Algorithm 1.

Feature Encryption: After obtaining the facial feature values, it is necessary to encrypt them using the generated keys. The entire facial floating-point feature values are converted into a vector called DoubleVector. The encryption

---

Algorithm 1: Key generation algorithm

Data: keypath
1 key_list ← HE.registration();
2 HE.save_key(keypath,key_list);
Result: public key, relinearized key, rotated key

---

is performed using the HE.encrypt function from the homomorphic encryption module. This function will return the encrypted ciphertext of the facial features. Subsequently, the local client uploads the encrypted facial feature ciphertext to the cloud-based ciphertext database. The algorithm flow can be seen in Algorithm 2.

Algorithm 2: Face Feature Encryption Algorithm

Data: feature
1 feature_dv ← DoubleVector(feature);
2 encrypted_feature ← HE.encrypt(feature_dv,context,publice_key);
Result: encrypted_feature

3.3 Ciphertext Feature Stream Distributed Computation

The approximate homomorphic encryption used in this paper is based on polynomial encryption, and only linear functions can perform corresponding approximate homomorphic operations. However, when the function model is nonlinear, it needs to undergo an approximate transformation to convert it into an approximate linear function.

Since face_recognition calculates the similarity between features using the Euclidean distance between them, the closer the Euclidean distance between two facial features, the higher the likelihood that the two faces belong to the same person. As shown below, the formula represents the Euclidean calculation in an N-dimensional space.

$$\text{dist}(\mathbf{x}, \mathbf{y}) = \sqrt{\sum_{i=1}^{n}(\mathbf{x}_i - \mathbf{y}_i)^2} \tag{1}$$

From the formula, it can be deduced that all operations are linear, except for the final square root calculation. In this paper, homomorphic computation is performed on the cloud side, and the homomorphic results are returned to the client, where the square root is performed in the plaintext domain.

The homomorphic encryption module currently does not directly support summation. Therefore, vector rotation is required to achieve summation with respect to the first element.

The approach proposed in this paper is based on the Flink data stream processing engine. Similar to traditional big data frameworks like Hadoop and Spark, Flink's parallel computing framework also supports the Map-Reduce paradigm. In Flink, data processing is done on data streams, where the Map operator performs operations on the data stream, and the Reduce operator aggregates the results of all data streams. Prior to the Reduce operation, a KeyBy operation is needed to split the data stream into different partitions and group data with the same key into the same partition.

In the proposed approach, we need to identify parallelizable steps in the ciphertext computation process. In the CKKS algorithm's ciphertext computation, ciphertexts are mutually independent and do not interfere with each other. Therefore, we can follow the parallelization techniques used in plaintext data processing to design parallelized computation for ciphertext data. The algorithm flow can be seen in Algorithm 3.

Algorithm 3: The Face Feature Distance Algorithm in a Data Stream Environment

---

    Data: cip1,cip_dict
1 gal_keys ← load_key(gal_keys_path);
2 **foreach** *key in encrypted_feature_dict* **do**
3   cip_sub ← HE.sub_cipher(cip1,cip_dict[key]);
4   cip_sqrt ← HE.square_cipher(cip_sub);
5   cip_rot ← Ciphertext(cipher_sqrt);
6   **foreach** *i = 1,2,...,k* **do**
7   evaluator.rotate_vector(cip_rot,1,gal_key);
8   cip_sum ← HE.add_cipher(cip_rot,cip_sqrt);
9   **end**
10   score_dict[key] = cip_sum;
11 **end**
    Result: score_dict

3.4   Feature Matching

The local client receives a dictionary of encrypted squared sums of facial feature values sent from the cloud. In this dictionary, the keys represent the user information identifiers, and the values represent the encrypted squared sums of facial feature values corresponding to each user. The HE.decrypt function is used to decrypt the values, and then a plaintext square root operation is performed locally, resulting in the plaintext facial feature values. Afterward, using the corresponding plaintext feature value threshold, the client filters out user information that meets the condition of having the smallest distance between facial feature values. The algorithm flow can be seen in Algorithm 4.

Algorithm 4: Feature Matching Algorithm

   Data: score_dict
1 **foreach** *key in score_dict* **do**
2 | score ← HE.decrypt(score_dict[key],secret_key);
3 | score_sqrt ← Math.sqrt(score);
4 | **if** *score_sqrt ≤ threshold && score_sqrt ≤ score_sqrt_min* **then**
5 | | score_sqrt_min = score_sqrt;
6 | | user_id = key;
7 | **end**
8 **end**
   Result: user_id

---

4 Algorithm Analysis

4.1 Security Analysis

Assuming that all entities (clients and servers) in the system model are honest and curious, they can honestly perform protocol computations but may try to obtain data from other entities. An adversary is defined with the following capabilities: (1) it may eavesdrop on the transmission of facial feature data; (2) it may eavesdrop and obtain intermediate results of facial feature template matching on the server, such as the squared Euclidean distance, leading to the inference of other private data based on the intermediate results and its own facial data.

Regarding the facial feature templates, the client's information is encrypted using homomorphic encryption before uploading it to the privacy service provider. The decryption key required for decryption can only be obtained by the client, and the server cannot decrypt the encrypted feature templates. This ensures that the server cannot access users' facial information.

The squared Euclidean distance, which is an intermediate result obtained through homomorphic operations, also requires the decryption key for decryption, and thus the server cannot obtain this data. Therefore, the client's data is secure and cannot be accessed by the server or the adversary, only requiring attention to the security of approximate homomorphic encryption itself. The hardness of the RLWE problem ensures the security of approximate homomorphic encryption.

4.2 Algorithm Performance Analysis

Serial Performance Analysis

We assume that the data stream contains N data points, each consisting of m features. After encryption, there are N ciphertexts, each ciphertext containing m ciphertext slots with data. The ciphertext computations are derived from two fundamental operations: homomorphic addition and homomorphic multiplication. As mentioned in the previous introduction to the CKKS algorithm, homomorphic addition and multiplication primarily involve additions and multiplications. In computer computations, the time consumed by addition and multiplication is roughly the same and denoted as T_m.

Homomorphic addition requires one addition operation and one modulo operation, while homomorphic multiplication requires 9 additions or multiplications and one modulo operation. Additionally, after each homomorphic multiplication, a re-scaling operation is needed, which requires 2 multiplications. The time consumed by the modulo operation is denoted as T_a.

In ciphertext computations, the time consumed by homomorphic addition or subtraction is denoted as $T_m + T_a$, and the time consumed by homomorphic multiplication is denoted as $11T_m + T_a$. Homomorphic vector dot product is implemented through operations that traverse ciphertext slots. Assuming the time taken to traverse one ciphertext slot is denoted as T_c, the time consumed by homomorphic vector dot product is $(10m + 2)T_m + T_a + mT_c$.

In this scheme, calculating the Euclidean distance requires N homomorphic vector subtractions, N homomorphic vector dot products, and N homomorphic additions. The time taken for homomorphic vector subtraction is $Nm(T_m + T_a)$, the time taken for homomorphic vector dot product is $N[(10m+2)T_m + T_a + mT_c]$, and the time taken for homomorphic addition is NT_m.

When performed sequentially, the time consumed for calculating the Euclidean distance is denoted as T_{serial}, and it is equal to the sum of the times taken for vector subtraction, vector dot product, and addition: $T_{serial} = Nm(T_m + T_a) + N[(10m + 2)T_m + T_a + mT_c] + NT_m = (11m + 3)NT_m + (m + 1)NT_a + NmT_c$.

Parallel Performance Analysis
Assuming parallelism, the data stream is divided into s sub-data streams and assigned to s maps for ciphertext computations. Each sub-data stream contains s data points, so it satisfies $k = N/s$. Assuming the cluster has d nodes, and on average, each node can process v sub-data streams, then $s = dv$.

The parallel ciphertext computation process can be divided into two parts: Map and Reduce. In the Map phase, data stream computations and communication between Maps are performed. In the Reduce phase, node scheduling and merging/sorting of sub-data streams are carried out. Let the time consumed by a single Map be M_0, and the total time for the Map phase be M. Then:

$$M = vM_0$$
$$M_0^{main} = (11m + 3)kT_m + (m + 1)kT_a + kmT_c$$

$$(2)$$

When there are s Maps working in the cluster, there will be at least $2s$ communication events. Let the communication time be $T_f = \delta sT_m$. The speedup of the algorithm in the Map phase can be calculated as follows:

$$\frac{T_{main}}{T_{Map}} \approx \frac{T_{main}}{vM_0^{main} + T_f}$$

$$= \frac{(11m+3)NT_m + (m+1)NT_a + NmT_c}{v[(11m+3)kT_m + (m+1)kT_a + kmT_c] + \delta sT_m}$$

$$= \frac{1}{\dfrac{kv}{N} + \dfrac{\delta sT_m}{(11m+3)NT_m + (m+1)NT_a + NmT_c}} \tag{3}$$

$$= \frac{d}{1 + \dfrac{d\delta ST_m}{(11m+3)NT_m + (m+1)NT_a + NmT_c}} \approx d$$

In practical environments, N is usually much larger than the number of cluster nodes and the number of sub-data streams, and the communication time between nodes is almost negligible. Therefore, the theoretical speedup in the Map phase can reach the theoretical value of d.

In the Reduce phase, the main time-consuming tasks are the integration and sorting of computation results from Map nodes and communication among nodes. Let the communication time in this process be $T_{fr} = \delta_1 sT_m$. Additionally, using a sorting algorithm with a complexity of $O(slogs)$, the sorting time is denoted as $S = \varepsilon slogsT_m$.

The overall speedup in the parallel computation process can be calculated as follows:

$$\frac{T_{main}}{T_P} = \frac{T_{main}}{vM_0^{main} + T_f + S + T_{fr}} \approx \frac{T_{main}}{vM_0^{main} + T_f}$$
$$\approx \frac{T_{main}}{T_{Map}} \approx d \tag{4}$$

In the experimental environment of this paper, the average computation time for the Map phase is 0.5121 s, while the average computation time for the Reduce phase is 0.6163 s. The Reduce phase's computation time is 1.204 times that of the Map phase.

$$S + T_{fr} \approx 1.2(vM_0^{main} + T_f) \tag{5}$$

The overall speedup is given by:

$$\frac{T_{main}}{T_P} = \frac{T_{main}}{vM_0^{main} + T_f + S + T_{fr}}$$
$$\approx \frac{T_{main}}{vM_0^{main} + T_f + 1.2(vM_0^{main} + T_f)} \approx \frac{1}{2.2}d \tag{6}$$

Based on the above analysis, it can be concluded that due to the relatively large computation time in the Reduce phase, the overall speedup will be approximately $1/2.2$ times the numwber of cluster nodes.

5   Experiments

Precision Comparison: In this experiment, the Labeled Faces in the Wild (LFW) dataset was used, which consists of 5,749 folders containing 13,233 images. The face verification task was performed using the face pairs information provided in pairs.txt, which includes 3,000 matching pairs and 3,000 non-matching pairs. The goal was to determine the facial similarity and verify whether two face images belonged to the same person. The CPU device used in this experiment is: 11th Gen Intel Core i7-11800H @ 2.30 GHz, Octa-Core. The memory device used is: 32 GB (Kingston DDR4 3200MHz). The Flink cluster was set up using Ubuntu virtual machines, consisting of one Taskmanager and two Jobmanagers.

The results are shown in Fig. 2.

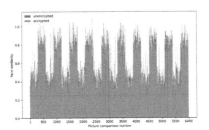

Fig. 2. Face Similarity

Obtain the matching results and conduct error analysis of the similarity between the unencrypted and encrypted schemes. The calculation of the error rate for the encryption scheme is done using the following formula:

$$error\_rate = \frac{ABS(Unencrypted - Encrypted)}{Unencrypted} * 100 \qquad (7)$$

Figure 3 corresponds to the error rate for 6,000 pairs of face verification, and the final accuracy was 98.05%, with an average error rate of 0.004618455%. The face verification scheme based on homomorphic encryption produces similarity calculation values that are close to the values obtained in plaintext. Furthermore, when compared with the accuracy results from [23], as shown in the Table 1, our method significantly improves accuracy and demonstrates higher efficiency.

Further experimental comparison of unencrypted and encrypted data is shown in Fig. 4. From the ROC curves of the False Acceptance Rate (FAR) and True Acceptance Rate (TAR), it can be observed that both unencrypted and encrypted ROC values are 98.3401%. The parallel processing of homomorphic encryption hardly affects the recognition accuracy, and this experimental result is consistent with the error analysis results.

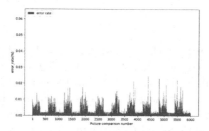

Fig. 3. Face Feature Error Rate

Table 1. Accuracy Comparison

Method	[24]	[25]	[23]	[ours]
Accuracy of the top-1 match	91.55%	95.50%	96.60%	98.05%
Feature volume	1600bits	3000bits	256bits	128bits

The smaller the values of False Acceptance Rate (FAR) and False Rejection Rate (FRR), the better the performance. However, changes in individual metrics can affect other metrics. From the Detection Error Tradeoff (DET) curve of FAR and FRR, it can be observed that the encrypted ERR (Equal Error Rate) value is 0.02567. A curve that leans towards the lower-left corner indicates better performance of the scheme, meaning that the difference between actual and measured values is small. The results are shown in Fig. 5.

Efficiency Comparison: This paper conducts serial and parallel tests on encrypted face recognition using a test dataset containing 10,000 pairs of encrypted face images. The parallelism of the cluster is adjusted by controlling the available slot slots and task parallelism. The experiment records the training time of the algorithm in serial and with different degrees of parallelism in the cluster. By using the serial time as a reference, the speedup ratio for different degrees of parallelism is calculated, and finally, the experimental results are analyzed. The results are shown in Table 2 and Fig. 6.

Based on the analysis of Table 2 and Fig. 6, the following conclusions can be drawn regarding the increase in available slots and parallelism of the Flink cluster:

(1) The overall execution time of the ciphertext gradually decreases with the increase in parallelism. Particularly, before reaching the maximum available slots in the cluster, the execution time of the ciphertext significantly decreases, leading to a notable improvement in algorithm performance. Once the parallelism exceeds the maximum available slots in the cluster, all nodes are fully utilized for computation. Moreover, the impact of increased parallelism on the cluster performance becomes relatively small, and the ciphertext execution time stabilizes gradually.

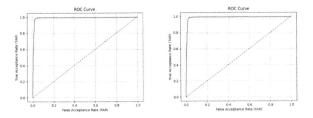

Fig. 4. ROC Curve for Encrypted and Unencrypted Data

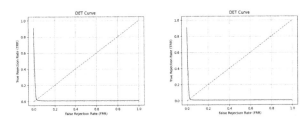

Fig. 5. DET Curve for Encrypted and Unencrypted Data

(2) As the size of the Flink cluster increases, theoretically, it can handle more data volume and more complex computational tasks. However, in practical applications, the extent of efficiency improvement does not always directly correlate with the size of the cluster. This is mainly due to the following factors:

- Data Partitioning: For some datasets, it may be challenging to distribute them evenly among all nodes in the Flink cluster. This leads to some nodes being overloaded while others remain idle, which decreases the overall efficiency of the cluster.
- Network Communication: With the growth of the cluster, the amount of data that needs to be transferred between nodes also increases. If the network bandwidth is insufficient or there are high network latencies, it can slow down the communication between nodes, thus affecting the overall efficiency of the cluster.
- Hardware Limitations: Before increasing the size of the Flink cluster, it is essential to ensure that each node has sufficient hardware resources such as memory, CPU, and disk space. Otherwise, adding more nodes may result in wastage of resources without a corresponding increase in efficiency.

(3) Due to the significant time consumption in the Reduce stage, the overall speedup ratio decreases. The highest overall speedup ratio is 4.8, which is close to $9/2.2 = 4.09$, and it is consistent with the theoretical analysis value mentioned earlier. Moreover, compared [23], the average encryption matching time per pair of images has reduced from 0.71 s to 63.5 ms, greatly improving the efficiency.

Table 2. Runtime and Speedup at Different Parallelism Levels

Parallelism	Execution time/ms	Speedup
1	3032227	1.00
2	1581397	1.91
3	1081455	2.80
4	898669	3.37
5	751976	4.03
6	690305	4.39
7	656067	4.62
8	643639	4.71
9	635354	4.77
10	632041	4.80
11	632626	4.79
12	636156	4.77

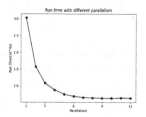 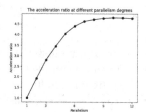

Fig. 6. Runtime and Speedup Curves

Feasibility Analysis: Based on the precision comparison experiments conducted on encrypted data, it can be concluded that the average error rate of the face recognition algorithm in the current encrypted domain is approximately 0.004618455%, which ensures high accuracy in practical application scenarios.

Furthermore, based on the efficiency comparison experiments, in the experimental environment with nine slots, the highest parallel efficiency for retrieving 10,000 facial images took approximately 635,354 ms. The average retrieval time for each image in the database was 63.5 ms. According to the conclusions from the performance analysis of parallel algorithms, by increasing the slots to 54, the acceleration ratio can reach approximately 24.5 times, and the retrieval time for each image is around 10.58 ms. If the encrypted clustering search algorithm is also applied simultaneously, it can effectively meet the requirements for daily applications.

6   Conclusion

This paper proposes a parallel privacy-preserving homomorphic encryption face verification scheme to address the issues of face information leakage and low computational efficiency in traditional face recognition. The scheme reconstructs the feature template matching protocol based on the homomorphic encryption module to achieve feature vector matching in the ciphertext domain. Security analysis and experimental comparisons demonstrate that the proposed scheme can achieve vector feature template matching while ensuring the security of the feature templates.

Leveraging the parallel computing concept of Map-Reduce, the scheme combines the stream data processing engine Flink with the homomorphic encryption algorithm CKKS to accelerate feature value matching in the ciphertext domain. Through algorithm performance analysis and experimental results, the acceleration ratio and accuracy of the algorithm are verified.

In conclusion, the proposed approach achieves efficient and accurate face verification while ensuring the security of data information during transmission and storage in the cloud.

References

1. Cheon, J.H., Kim, A., Kim, M., Song, Y.: Homomorphic encryption for arithmetic of approximate numbers. In: Takagi, T., Peyrin, T. (eds.) ASIACRYPT 2017. LNCS, vol. 10624, pp. 409–437. Springer, Cham (2017). https://doi.org/10.1007/978-3-319-70694-8_15
2. Titus, A.J., Kishore, S., Stavish, T., Rogers, S.M., Ni, K.: Pyseal: a python wrapper implementation of the seal homomorphic encryption library (2018)
3. Li, Y., Ng, K.S., Purcell, M.: A tutorial introduction to lattice-based cryptography and homomorphic encryption (2022)
4. Dowerah, U., Krishnaswamy, S.: Towards an efficient LWE-based fully homomorphic encryption scheme. IET Inf. Secur. **16**(4), 16 (2022)
5. Xiaoming D., Department, E.T.: Research on fully homomorphic encryption schems. Electronics World (2016)
6. Zhang, Z., Cheng, P., Chen, J., Wu, J.: Secure state estimation using hybrid homomorphic encryption scheme. IEEE Trans. Control Syst. Technol. **29**, 1704–1720 (2020)
7. Wang, Y., Liang, X., Hei, X., Ji, W., Zhu, L.: Deep learning data privacy protection based on homomorphic encryption in aiot. Mob. Inf. Syst. **2021**(2), 1–11 (2021)
8. Park, J., Kim, D.S., Lim, H.: Privacy-preserving reinforcement learning using homomorphic encryption in cloud computing infrastructures. IEEE Access **8**, 203564–203579 (2020)
9. Aloufi, A., Hu, P., Song, Y., Lauter, K.: Computing blindfolded on data homomorphically encrypted under multiple keys: a survey. ACM Comput. Surv. (CSUR) **54**, 1–37 (2021)
10. Yang, X., Yi, X., Kelarev, A., Han, F., Luo, J.: A distributed networked system for secure publicly verifiable self-tallying online voting. Inf. Sci. **543**, 125–142 (2021)
11. Xu, W., Wang, B., Hu, Y., Duan, P., Zhang, B., Liu, M.: Multi-key fully homomorphic encryption from additive homomorphism. Comput. J. **66**(1), 197–207 (2023)

12. Fang, H., Qian, Q., Chen, M.L.: Privacy preserving machine learning with homomorphic encryption and federated learning. Future Internet **13**, 94 (2021)

13. Wibawa, F., Ozgur Catak, F., Sarp, S., Kuzlu, M., Cali, U.: Homomorphic encryption and federated learning based privacy-preserving cnn training: Covid-19 detection use-case. arXiv e-prints (2022)

14. Zhang, J., Jiang, Z.L., Li, P., Yiu, S.M.: Privacy-preserving multikey computing framework for encrypted data in the cloud. Inf. Sci. **575**, 217–230 (2021)

15. Park, J.H.: Homomorphic encryption based privacy-preservation for iomt. Appl. Sci. **11**, 8757 (2021)

16. Mohammed, S., Basheer, D.: From cloud computing security towards homomorphic encryption: a comprehensive review. TELKOMNIKA (Telecommunication Computing Electronics and Control) (2021)

17. Gentry, C., Halevi, S., Smart, N.P.: Better bootstrapping in fully homomorphic encryption. In: Fischlin, M., Buchmann, J., Manulis, M. (eds.) PKC 2012. LNCS, vol. 7293, pp. 1–16. Springer, Heidelberg (2012). https://doi.org/10.1007/978-3-642-30057-8_1

18. Gentry, C., Halevi, S.: Implementing gentry's fully-homomorphic encryption scheme. In: Paterson, K.G. (ed.) EUROCRYPT 2011. LNCS, vol. 6632, pp. 129–148. Springer, Heidelberg (2011). https://doi.org/10.1007/978-3-642-20465-4_9

19. Tibouchi, M.: Fully homomorphic encryption over the integers: from theory to practice. NTT Techn. Rev. **12**(7), 273–81 (2014)

20. Zhao, D.: Rache: radix-additive caching for homomorphic encryption (2022)

21. Chillotti, I., Gama, N., Georgieva, M., Izabachène, M.: Faster fully homomorphic encryption: bootstrapping in less than 0.1 seconds. In: Cheon, J.H., Takagi, T. (eds.) ASIACRYPT 2016. LNCS, vol. 10031, pp. 3–33. Springer, Heidelberg (2016). https://doi.org/10.1007/978-3-662-53887-6_1

22. Ertaul, L.: Implementation of homomorphic encryption schemes for secure packet forwarding in mobile ad hoc networks (manets) (2022)

23. Ma, Y., Wu, L., Gu, X., He, J., Yang, Z.: A secure face-verification scheme based on homomorphic encryption and deep neural networks. IEEE Access **5**, 16532–16538 (2017)

24. Jin, X., Liu, Y., Li, X., Zhao, G., Guo, K.: Privacy preserving face identification in the cloud through sparse representation. In: Chinese Conference on Biometric Recognition (2015)

25. Osadchy, M., Pinkas, B., Jarrous, A., Moskovich, B.: Scifi - a system for secure face identification. In: IEEE Symposium on Security & Privacy (2010)

A Graph-Based Vertical Federation Broad Learning System

Junrong Ge, Fengyin Li(✉) ⓘD, Xiaojiao Wang, Zhihao Song, and Liangna Sun

School of Computer Science, Qufu Normal University, Rizhao, China
`lfyin318@qfnu.edu.cn`

Abstract. A broad learning system is a lightweight deep neural network with breadth expansion, which is widely used in face recognition, error detection, and so on. The broad learning system can make full use of grid data, but it is not suitable for utilizing graph data that can represent data relationships. Due to the emphasis on data privacy, the feature data of graph for training models is often in a fragmented state. Based on the above problems, this paper introduces the vertical federation idea and the graph convolutional neural network into the broad learning system and uses the graph neural network to assist the broad learning system in extracting features of graphs. We use the isolated graph information jointly extracted by the vertical federation framework for broad learning and propose a graph-based vertical federation broad learning system. Since the weights for extracting features are randomly generated during the initial graph establishment phase. There is no guidance for extracting features, and the quality of the extracted features is not guaranteed. Therefore, this paper introduces the extreme learning auto-encoder into the graph-based vertical federation broad learning system to generate weights for extracting graph features.

Keywords: broad learning system · graph neural network · vertical federation

1 Introduction

A broad learning system is a simple deep neural network proposed by Professor Chen Junlong in 2018 [1]. The broad learning system is not only simple in structure but also does not need to be extended in depth. It does not involve complex hyperparameter calculations, and the training speed is fast [2]. In addition, when the effect of the wide learning system does not meet the requirements, the incremental learning algorithm can be used to quickly calculate the model without retraining [3]. In the field of security protection, the broad learning network improves the scene adaptability of the face recognition algorithm in complex scenes and improves the reliability of artificial intelligence recognition [4]. In addition, the algorithm model built by the broad learning network can solve the problem of algorithm labeling errors. Broad learning leverages the weights of different parts of the graph-regularized sparse autoencoder model for fine-tuning,

© The Author(s), under exclusive license to Springer Nature Singapore Pte Ltd. 2024
J. Vaidya et al. (Eds.): AIS&P 2023, LNCS 14510, pp. 53–62, 2024.
https://doi.org/10.1007/978-981-99-9788-6_5

maintaining the input manifold, while enhancing the stability of broad learning and improving hyperspectral image analysis capabilities [5]. Broad learning systems are also widely used in image classification, numerical regression, EEG signals, and other fields [6, 7].

With the advent of the era of big data, the value of data is not only reflected in the information contained in itself but also in the relationship between data [10]. Therefore, it becomes extremely important to fully mine the information of the relationship between the data. Graphs can represent rich relationships between data [11]. The research on graphs is in full swing and has been involved in the fields of natural language processing, computer vision, biomedicine, industrial recommendation, and industrial risk control [12–14].

Broad can fully extract information from grid data but is not suitable for mining the information of graphs. Due to the emphasis on data privacy and value, the government, institutions, and individuals have strengthened the control of data, making the feature data of the graph in a fragmented state [15,16]. Vertical federated learning provides a solution to the above problems [17,18]. Federated learning is similar to a distributed learning framework with privacy-preserving, which provides the possibility for joint training of intelligent models in the state of feature data islands [19]. Federated learning is involved in the Internet of Things, which can jointly utilize the information of edge devices, which is conducive to the expansion of the scope of the Internet of Things [20]. In the medical field, federated learning integrates diagnostic information from various places to realize intelligent disease diagnosis and analysis [21,22].

To solve the problem that the broad learning system is not suitable for utilizing isolated graph feature data, a graph-based vertical federation broad learning system is proposed. The contributions of this paper are as follows:

1 This paper introduces the idea of vertical federation into the broad learning system, uses the graph neural network to assist the feature extraction of the broad learning system, and proposes a graph-based vertical federation broad learning system.
2 In this paper, the extreme learning auto-encoder is used to generate weights to replace randomly generated weights, assisting the feature extraction of the broad learning system.

2   Preliminaries

In this section, we give the preliminaries of this paper, of which Sect. 2.1 is the broad learning system, Sect. 2.2 is the vertical federated learning, and Sect. 2.3 is the extreme learning auto-encoder.

2.1   Broad Learning System

The broad learning system is a deep neural network proposed based on the random vector functional-linked neural network [1]. Its training process is as follows.

a. Preprocess the data by normalization.
b. Generate mapped features and enhancement nodes
   Each set of mapped features is generated as shown in Eq. (1), as in Fig. 1.

$$Z_i = \varphi(Xw_i + \beta_i), i = 1, 2, ...n \tag{1}$$

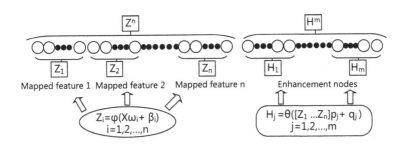

Fig. 1. Mapped features and enhancement nodes generation diagram

First, X is multiplied by the weight matrix w_i and then added to the bias matrix β_i, and finally activated by the activation function $\varphi()$ to obtain the mapped features. The n groups of enhanced features are horizontally spliced and expressed as Z^n. The generation of each group of enhanced nodes is shown in formula (2).

$$H_j = \theta(Z^n p_j + q_j), j = 1, 2, ...m \tag{2}$$

Z^n is multiplied by the weight matrix p_j and then added to the bias matrix q_j to obtain the jth enhanced node H_j through the activation function. M groups of enhanced nodes are spliced horizontally to obtain H^m. Among them, the weight matrix and bias matrix are randomly generated, once generated, they will not be changed. The matrix A is obtained by splicing m groups of mapping features and n groups of increasing features, $A=[Z^n|H^m]$.

c. Compute the parameters of a broad learning system

In this step, the pseudo-inverse of A and the output connection weight of the broad learning system are computed.
A broad learning system can be expressed as Eq. (3).

$$Y = [Z^n|H^m]\,W = AW \tag{3}$$

where W is the output connection weight of the model. According to (3), we can see (4).

$$W = A^+Y = [Z^n|H^m]^+Y \tag{4}$$

A^+ is the pseudo-inverse of A. The loss function of the broad learning system can be expressed as the following optimization problem.

$$\underset{W}{\arg\min} \|AW - Y\|_v^{w1} + t\|W\|_u^{w2} \tag{5}$$

When $w1 = w2 = v = u = 2$, the above problem is transformed into a l_2 norm problem. The t is a coefficient of the sum of squares of W, expressing a constraint on the sum of squares of the weight.

$$A^+ = (tI + A^T A)^{-1} A^T \tag{6}$$

W can be quickly calculated according to Eq. (7).

$$W = \left(tI + A^T A\right)^{-1} A^T Y \tag{7}$$

d. Test the accuracy.

Test the accuracy of the broad learning system. If the accuracy rate does not meet the requirements, the broad learning system adds nodes to improve accuracy until it terminates when the accuracy meets the requirements.

2.2 Vertical Federated Learning

Vertical federated learning is a kind of federated learning, which is based on data sets with the same sample space and different feature spaces [17]. That is, vertical federated learning can be understood as federated learning divided by features [18], as in Fig. 2.

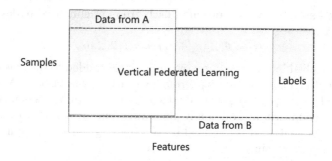

Fig. 2. Vertical federal scenario diagram

The training process of the vertical federated learning model introduces a third party to preserve the privacy of the participants. It helps the clients to conduct secure federated learning. The third party collects the intermediate results to calculate the gradient and loss value and forwards the computed results to the corresponding clients. The data received by the third party is encrypted or obfuscated, so the data is secure [19].

2.3 Extreme Learning Auto-Encoder

Extreme learning auto-encoder is an unsupervised learning method [8]. It sets the desired output as the input data. It computes the connection weight between the hidden and output layers by minimizing the reconstruction error [9]. Given the training set X, the hidden layer of the extreme learning auto-encoder can be expressed by Eq. (8).

$$H = p(XW + \beta) \tag{8}$$

Where p is an activation function, and the weights W and β are randomly generated, but satisfy $W^T W = I$ and $\beta\beta^T = I$. The loss function of the extreme learning auto-encoder is defined as Eq. (9).

$$O_{ELM-AE} = \tfrac{1}{2}||HW' - X||_F^2 + \tfrac{t}{2}||W'||_F^2 \tag{9}$$

$||\ \ ||_F$ represents the Frobenius norm; W' is the output connection weight of the auto-encoder and t is a predefined parameter. Deriving the loss function and setting it equal to 0 yields Eq. (10).

$$\nabla O_{ELM-AE} = H^T HW' + tW' - H^T X = 0 \tag{10}$$

W' can be calculated by Eq. (11).

$$W' = \begin{cases} (H^T H + tI)^{-1} H^T X & N \geq l \\ H^T (tI + HH^T)^{-1} X & N < l \end{cases} \tag{11}$$

Among them, N represents the number of samples, and l is the number of hidden layer nodes.

3 Graph-Based Vertical Federation Broad Learning System

In order to solve the learning problem of the broad learning system in the state of the feature data of graph islands, this paper first uses vertical federation to cooperate with clients holding features of graph data for broad learning. There are two types of entities in the graph-based vertical federation broad learning system, the client and the third party. Among them, clients hold isolated graphs, and different clients hold the same data samples, with different features. The clients are responsible for extracting features from local data. The third party is responsible for the calculation of parameters of the graph-based vertical federation broad learning system. In the graph-based vertical federation broad learning system, the local data held by clients is graphs, so clients need to extract the features of the local data. This paper assumes that there are k clients who hold the graph for training the model, and the kth cleint holds the label data, as shown in Fig. 3. The following paper will be divided into two parts: the feature extraction and the graph-based vertical federation broad learning system.

3.1 The Feature Extraction

Clients extract features from local data by establishing a graph neural network to provide data for the graph-based vertical federation broad learning system.

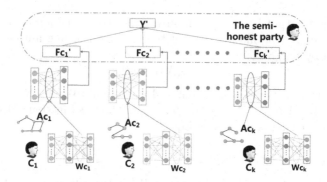

Fig. 3. Graph-based vertical federation broad learning system

Firstly, clients use the extreme learning auto-encoder to generate weights for the feature extraction of the graph. The following is the specific process of extraction.

The client C_i performs nonlinear transformation on local data to generate a hidden layer for extreme learning auto-encoder. The output of the hidden layer H_{Ci} is obtained by multiplying X_{Ci} with the weight W and adding β, and then activating the activation function. Among them, X_{Ci} is local data, W and β are randomly generated. At this time, the extreme learning auto-encoder can be expressed as $H_{Ci}W_{Ci} = X_{Ci}$. W_{Ci} is the connection weight of the extreme learning auto-encoder. In order to prevent overfitting, the loss function of the extreme learning auto-encoder is added a penalty term $\|W_{Ci}\|_F$. The output connection weights W_{Ci} of the extreme learning auto-encoder are obtained according to Eqs. (9)–(11).

The common form of graph feature extraction is $H = f(\bar{A}XW)$. Among them, X is the sample data, $\bar{A} = D'^{-\frac{1}{2}}A'D'^{-\frac{1}{2}}$, $A' = A + I$. A is the adjacency matrix of the graph, D' is the degree matrix of A', I is the identity matrix, and $f()$ is the activation function. W weight matrix, which is randomly generated. In the graph-based vertical federation broad learning system, the operation of the graph feature extraction on clients is as follows.

Client C_i uses Eq. (12) to extract features from local data to get F_{Ci}, $i \in [1, k]$. Among them, X_{Ci} is the local data of the client. $A_{Ci} = D_i^{-\frac{1}{2}}A_iD_i^{-\frac{1}{2}}$, $A_i = A^i + I$. D_i is the degree matrix of A_i. I is the identity matrix. A^i is the adjacency matrix of the local graph and W_{Ci} is the weight obtained by the extreme learning auto-encoder.

$$Fc_i = \phi(Ac_iXc_iWc_i), i = 1, 2, ..., k \tag{12}$$

3.2   Graph-Based Vertical Federation Broad Learning System

In the graph-based vertical federation broad learning system, clients use the output connection weight of the extreme learning auto-encoder to perform feature extraction on the local data. Client encrypts the extracted feature information and then transmits it to the third party. The third party receives the data from clients to calculate the weight of the broad learning system and transmits the calculation results to clients. The client decrypts the received data to complete the establishment of a graph-based vertical federation broad learning system. The establishment of the graph-based vertical federation broad learning system is as in Algorithm 1.

Algorithm 1. Graph-based vertical federation broad learning system

---

1: Clients C_1 to C_k generate weights $W_{C1'}$, W_{C2}', ..., W_{Ck}' by extreme learning auto-encoder according to the **Eq.** (8) and **Eq.** (11) .

2: C_1 to C_k perform feature extraction on local data according to **Eq.** (12) to obtain F_{C1}, F_{C2}, ..., F_{Ck}. C_1 to C_k encrypt F_{C1}, F_{C2}, ..., F_{Ck} into F_{C1}', F_{C2}', ..., F_{Ck}' and then forwards them to the third party.

3: The third party receives the data from the client and splices it into F_c.

4: The third party calculates the output connection weights W'' according to **Eq.** (6) and **Eq.** (7), tests the accuracy of the model.

5: The third party returns output connection weight W''.

6: The client restores the real output connection weight W of the graph-based vertical federation broad learning system.

The client C_i encrypts the result F_{Ci} of local feature data feature extraction into F_{Ci}' and forwards it to the third party. The third party receives the data F_{C1}', F_{C2}', ..., F_{Ck}' from k clients, and splices them horizontally into $F_c' = [F_{C1}'|F_{C2}'|...|F_{Ck}']$. At this time, the graph-based vertical federation broad learning system can be expressed as $F_c'W'' = Y'$, where Y' is the encrypted label. The third party sets the penalty parameter t, and calculates the output connection weight W'' according to (6) and (7). The third party transmits the encrypted W'' to clients. The clients get the real output connection weight W according to W''.

Since then the initialization of the model has ended. The third-party test accuracy, if the accuracy does not meet the requirements, the third party will notify the client to continue feature extraction until the accuracy meets the requirements.

4   Experiment

In this section, we performed experiments and gave experimental results to demonstrate the ideas proposed in this paper. The experiments in this paper were conducted on the Cora data set. In order to imitate the graph feature data for training the graph-based vertical federation broad learning system, which is held by different clients, this paper divides the feature data of Cora equally by column. The two parts of equally divided data are held by two clients respectively.

In the experiment, we set the number of features extracted by the client to be 50, 100, and 700 respectively. Since it is assumed that there are two customers in the model, the number of features in the model is 100, 200, and 1400 respectively. The experimental results are shown in Table 1.

From Table 1, when the number of features goes from 100 to 1400, the accuracy of the model increases from 0.572 to 0.990, an increase of 0.4. Therefore, as the number of extracted features increases, the effect of the model will continue to increase. Table 1 is visualized as Fig. 4.

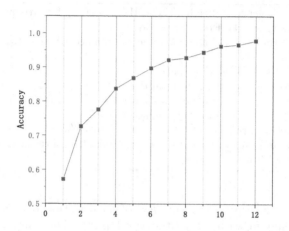

Fig. 4. The effect of the model under different feature extraction

Table 1. Table captions should be placed above the tables.

Extracted features	100	200	300	400	500	600	700	800	900	1000	1100	1200
Accuracy	0.572	0.727	0.776	0.837	0.868	0.897	0.922	0.928	0.944	0.962	0.966	0.978

5   Conclusion

The graph-based vertical federation broad learning system proposed in this paper can utilize the isolated graph data, which provides a method for the broad learning of graphs in vertical federated scenarios and expands the application range of the broad learning system. The graph-based vertical federation broad learning system uses the extreme learning auto-encoder to guide the feature extraction of graphs, which facilitates the graph-based vertical federation broad learning system to fully extract data information.

In the future, we will work on improving the accuracy and security of the graph-based vertical federation broad learning system, and contribute to the research of broad learning system.

References

1. Chen, C.L.P., Liu, Z.: Broad learning system: an effective and efficient incremental learning system without the need for deep architecture. IEEE Trans. Neural Netw. Learn. Syst. **29**, 10–24 (2017)
2. Gong, X., Zhang, T., Chen, C.L.P., Liu, Z.: Research review for broad learning system: algorithms, theory, and applications. IEEE Trans. Cybern. **52**, 8922–8950 (2021)
3. Chen, C.L.P., Liu, Z., Feng, S.: Universal approximation capability of broad learning system and its structural variations. IEEE Trans. Neural Netw. Learn. Syst. **30**(4), 1191–1204 (2018)
4. Zhao, H., Zheng, J., Deng, W., Song, Y.: Semi-supervised broad learning system based on manifold regularization and broad network. IEEE Trans. Circ. Syst. **67**, 983–994 (2020)
5. Kong, Y., Wang, X., Cheng, Y., Chen, C.P.: Hyperspectral imagery classification based on semi-supervised broad learning system. Remote Sens. **10**(5), 685 (2018)
6. Li, Y., Li, T., Zuo, Y.: Port throughput forecasting based on broad learning system with considering influencing factors, pp. 4129–4134 (2020)
7. Zhao, H., Zheng, J., Xu, J., Deng, W.: Fault diagnosis method based on principal component analysis and broad learning system. IEEE Access **7**, 99263–99272 (2019)
8. Haidong, S., Hongkai, J., Xingqiu, L., Shuaipeng, W.: Intelligent fault diagnosis of rolling bearing using deep wavelet auto-encoder with extreme learning machine. Knowl.-Based Syst. **140**, 1–14 (2018)
9. Sun, K., Zhang, J., Zhang, C., Hu, J.: Generalized extreme learning machine autoencoder and a new deep neural network. Neurocomputing **230**, 374–381 (2017)
10. Duan, Y., Edwards, J.S., Dwivedi, Y.K.: Artificial intelligence for decision making in the era of Big Data-evolution, challenges and research agenda. Int. J. Inf. Manag. **48**, 63–71 (2019)
11. Zhou, J., et al.: Graph neural networks: a review of methods and applications. AI open **1**, 57–81 (2020)
12. Niepert, M., Ahmed, M., Kutzkov, K.: Learning convolutional neural networks for graphs. In: International Conference on Machine Learning, vol. 1, pp. 2014–2023 (2016)
13. Wu, Z., Pan, S., Chen, F., Long, G., Zhang, C., Philip, S.Y.: A comprehensive survey on graph neural networks. IEEE Trans. Neural Netw. Learn. Syst. **32**(1), 4–24 (2020)
14. Xu, K., Hu, W., Leskovec, J., Jegelka, S.: How powerful are graph neural networks. arXiv preprint arXiv:1810.00826 (2018)
15. Egorova, K.S., Toukach, P.V.: Glycoinformatics: bridging isolated islands in the sea of data. Angewandte Chemie Int. Ed. **57**, 14986–14990 (2018)
16. Terzi, D.S., Terzi, R., Sagiroglu, S.: A survey on security and privacy issues in big data. In: The 10th International Conference for Internet Technology and Secured Transactions (ICITST), pp. 202–207 (2015)
17. Yang, Q., Liu, Y., Chen, T., Tong, Y.: Federated machine learning: concept and applications. ACM Trans. Intell. Syst. Technol. (TIST) **10**(2), 1–19 (2019)
18. Liu, Y., et al.: Vertical federated learning. arXiv preprint arXiv:2211.12814 (2022)
19. Feng, S., Yu, H.: Multi-participant multi-class vertical federated learning. arXiv preprint arXiv:2001.11154 (2020)

20. Imteaj, A., Thakker, U., Wang, S., Li, J., Amini, M.H.: A survey on federated learning for resource-constrained IoT devices. IEEE Internet Things J. **9**, 1–24 (2021)
21. Rieke, N., et al.: The future of digital health with federated learning. NPJ Dig. Med. **3**, 119 (2020)
22. Sheller, M.J., et al.: Federated learning in medicine: facilitating multi-institutional collaborations without sharing patient data. Sci. Rep. **10**, 12598 (2020)

EPoLORE: Efficient and Privacy Preserved Logistic Regression Scheme

Wendan Zhang, Yuhong Sun$^{(\boxtimes)}$, Sucheng Yan, Hua Wang , Yining Liu, and Chen Zhang

Qufu Normal University, Qufu, China
sun_yuh@163.com

Abstract. Logistic regression, as one of the classification method, is widely used in machine learning. Due to the complexity of training process, outsourcing the training task to a third party is a feasible choice, while the plain or direct outsourcing will inevitabaly lead to privacy leakage. To address this problem, this paper proposes an efficient privacy-preserving outsourced logistic regression (EPoLORE) scheme. In securely training the model, we design related protocols: floating-point conversion, integer multiplication, vector inner product, and activation function based on a distributed double-trap public key cryptosystem (DT-PKC), allowing the cloud server to effectively perform the integer and floating-point computations with ciphertexts of training data. In such a way, the privacy of training data is preserved and the model can obtain the accuracy approximate to that of the regular model trained in plaintext. The Security of the protocols is analyzed, thereby demonstrating that EPoLORE meets the security requirements. The corresponding experiments show the effectiveness of the proposed scheme and the comparison of model accuracy with the regular training model.

Keywords: machine learning · logistic regression · privacy-preserving

1 Introduction

Artificial Intelligence (AI) has gradually entered into many fields of human society, and the machine learning(ML) supporting AI is developing rapidly. The logistic regression(LR), as one of the ML technology, is commonly used to solve data classification problems in many fields. For example, it is used to predict the risk of developing a certain disease based on the observed patient characteristics in medicine [1]. Another example is in politics, it is used to predict a person's voting behavior based on personal data such as age, income, gender, and previous voting, etc. [2]. LR is also used in finance to predict the likelihood of homeowners defaulting on mortgage loans or committing fraud in credit card transactions [3]. Same as other ML tools, LR requires sufficient training data to train models. Though the cloud computing is usually employed to train the mass data, the security has to be addressed. Therefore, it is necessary to design a logistic regression algorithm with privacy protection without compromising the validity of the data.

© The Author(s), under exclusive license to Springer Nature Singapore Pte Ltd. 2024
J. Vaidya et al. (Eds.): AIS&P 2023, LNCS 14510, pp. 63–77, 2024.
https://doi.org/10.1007/978-981-99-9788-6_6

1.1 Related Work

To address the privacy protection issues in ML, researchers have proposed a series of methods and techniques. Han et al. [4] proposed a LR model that can be trained on the encrypted data, which demonstrates the practical feasibility of the LR training on mass encrypted data for the first time. Mandal et al. [5] proposed PrivFL, which achieves a robust and secure training process through the iterative execution of a secure multi-party global gradient update protocol. P. Mohassel et al. [6] proposed an efficient privacy-preserving protocol for training data in LR. It is considerably faster to implement compared to existed privacy-preserving protocols. J. H. Cheonet al. [7] explored privacy protection techniques that reduces the number of gradient descent iterations. Shi et al. [8] built a grid LR framework based on secure multi-party computation that protects not only the end data but also all intermediate data exchanged during the learning phase. Li et al. [9] proposed a new differential privacy outsourcing scheme that allows providers to go offline after uploading a dataset, thus achieving lower communication costs. Zhu et al. [10] proposed a privacy-preserving ML training framework called Heda based on Paillier and RSA homomorphic encryption, which achieves the effectiveness of privacy-preserving ML training without losing model accuracy. Aono et al. [11] proposed a secure system for protecting both the training and predicting data in logistic regression via homomorphic encryption, and the system is very scalable. Sergiu Carpov et al. [12] proposed TFHE and HEAAN based on the hybrid framework Chimera that allows for switching between different families of fully homomorphic schemes. But this scheme requires a high computational cost. Similarly, CKKS [13], as the most popular FHE scheme, has the drawback of an imprecise decryption result, and it is not popular in training.

In order to guarantee the accuracy and privacy, we propose an efficient privacy-preserving outsourced logistic regression training scheme (EPoLORE) used in the cloud computing. With the designed protocols of privacy-preserving integer multiplication, vector inner product, and activation function, this scheme achieves the privacy-preserving of training data and accuracy of classification. We analyze and evaluate the system for privacy preserving training and logistic regression models. Experiments show the feasibility of the protocols.

1.2 Contribution

A summary of our contributions is as follows.

(1) We designed a floating-point conversion algorithm using finite scaling method to convert the floating-point number used in ML to integers for cryptosystem.
(2) We proposed three computing protocols for integer or floating-point number: multiplication, vector inner product and the activation function.
(3) We proved the security of the protocols, against semi-honest adversary in the simulation paradigm.
(4) We implemented and evaluated the efficiency of EPoLORE for training LR models on Iris and Heart datasets.

2   Preliminaries and System Model

In this section, we describe the DT-PKC cryptosystem and LR algorithm used in this paper. In addition, we show the system model and the security goals.

2.1   Distributed Two Trapdoors Public-Key Cryptosystem (DT-PKC)

DT-PKC is an efficient privacy-preserving outsourced computing toolkit, proposed by Liu et al. [14]. In DT-PKC, the system master decryption key (MK) is split into two parts, and neither part can decrypt a ciphertext separately. This resolves the security issue that arises when the central node possesses the master key. The DT-PKC consists the algorithms of Initialize, Key generation (Key-Gen), Encryption(Enc), Decryption With Weak Private Key, Decryption With Strong Private Key, Strong Private Key Splitting, Partial Decryption With Partial Strong Private Key StepOne (PDO) and Partial Decryption With Partial Strong Private Key StepTwo (PDT).

DT-PKC has the homomorphic property. Give two ciphertexts $[x_1]$ and $[x_2]$ under the same public key, the additive homomorphic operation is defined as $[x_1 + x_2] = [x_1] \cdot [x_2]$. Liu et al. [14] provides proofs of correctness and semantic security of the DT-PKC.

2.2   Logistic Regression

Logistic Regression (LR) is used to establish the logical relationship between data and its labels. LR predicts the label of the original data using logical operations and calculates the degree of influence of each data item on the output. The prediction is determined by the sigmoid function:

$$S(z) = \frac{1}{1 + e^{-z}} \tag{1}$$

The LR hypothesis function is defined as follows:

$$h_\theta(x) = S(w^T x) \tag{2}$$

Here, x is input and w is the feature vector coefficient we are looking for.

We choose to use Random Gradient Ascent (RGA) to obtain the parameter w that maximizes the likelihood function, for RGA only requires one sample point when updating regression coefficients, while the gradient descent algorithm requires iterating through the entire dataset with each coefficient update.

2.3   System Model

The system mainly consists of Key Generation Center (KGC), Data Provider (DP), Cloud Storage Server (CSS), Crypto-Service provider (CSP), Service Forwarder (SF) and Service Requster (SR), as shown in Fig. 1.

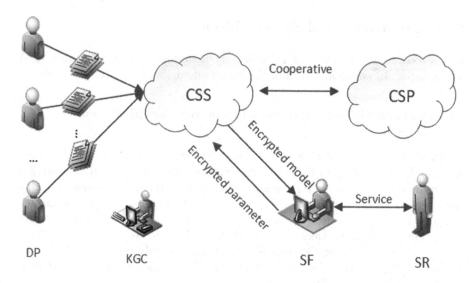

Fig. 1. System model.

- KGC: The KGC generates and distributes the public/private key pairs for all the participants. Then, it may be offline.
- DP: DP provides encrypted data to the CSS for training.
- CSS: CSS provides storage and computing service to the DPs, and it completes the training process with the CSP.
- CSP: CSP assists CSS in training the model.
- SF: SF provides prediction service after receiving the encrypted models. In addition, when SF receives a service request, it will forward it to CSS.
- SR: SR is the user who sends the encrypted data to the SF for secure classification.

Entities in the system model are classified as trusted and semi-honest [17] based on whether other entities trust them. KGC is considered to be fully trusted, while DP, CSS, CSP, SF, and SR are all assumed to be semi-honest. In addition, we assume that the CSP will not collude with the CSS.

2.4   Security Goals

The security goals of the proposed scheme should meet the following requirements.

- Secure outsourced storage: DPs outsource their data to the cloud servers for operation without revealing privacy.
- Secure computation: All protocols including multiplication, inner vector and the sigmoid function performed on ciphertext using homomorphic operations, do not leak the original data.
- Secure communication: The DP_k does not know the information of other $DP_j(j \neq k, j = 1, \ldots, n)$, and no DP can know the final model parameters.

2.5 Notations Description

Assume there are n data providers, that is $DP_k, k = 1, \ldots, n$ in the system. Each DP holds his own keys of PK_k/SK_k. The master key for decryption is splitted as $\lambda_i(i \in 1, 2)$. Unless otherwise specified, $[x]_{PK}$ is used to represent the ciphertext encrypted with PK.

3 The Proposed Scheme

In this section, we present the design idea, the processing of floating-point numbers, and the protocols designed for EPoLORE.

3.1 Design Idea

Although most of the ML algorithms are highly complicated, they are the recombination and iteration of the basic operations such as addition, multiplication, and so on. With the investigation on the typical supervised ML algorithms, we found that the basic operations in training are addition, subtraction, multiplication, and exponentiation. The additive homomorphic encryption can perform basic addition and subtraction operations, but there is currently no direct method for multiplying ciphertexts or performing exponentiation operations. Therefore, it is necessary to design secure computing protocols for these fundamental operations.

3.2 Floating-Point Number Conversion

Most data in samples and parameters in model are floating-point number. However, homomorphic encryption scheme cannot perform operations on floating-point number directly. To address this problem, we have to convert the floating-point number to integers firstly. The conversion algorithm is shown in Algorithm 1.

Algorithm 1. Secure Float-Point Format Conversion
___
Input: Floating-point x , precision $E[E < L(N)]$
Output: Integer-point x^*
1. If $x > 0$
2.   then $x^* = x \cdot 2^E$
3. else
4.   then $x^* = N - x \cdot 2^E$
5. end if
6. Return x^*

For example, a floating-point number $x = 3.1$ and a fixed precision $E = 10$, we can convert x into an integer $x^* = 3.1 \times 2^{10} = 3174$. Similarly, the floating-point number 3.1 can be obtained by $3.1 = 3174 \div 2^{10}$.

3.3 Secure Integer Multiplication (SIM)

The goal of SIM is to input $[x]_{PK_1}$, $[y]_{PK_2}$ and compute the ciphertext $[z]_{PK_3} = [x \cdot y]_{PK_3}$ under the new key PK_3 without revealing any plaintext. The calculation process is completed by CSS and CSP.

- CSS firstly chooses four random numbers r_x, r_y, R_x, R_y from the group Z_N, and encrypts them with PK_1 and PK_2 by DT-PKC.Enc to get $[r_x]_{PK_1}$, $[r_y]_{PK_2}, [R_x]_{PK_1}, [R_y]_{PK_2}$, where PK_1 and PK_2 is the public key of the CSS and CSP, respectively. Then it computes the ciphertext $[x']_{PK_1}, [y']_{PK_1}$, $[S]_{PK_1}, [T]_{PK_2}$ by homomorphism.
  Next, DT-PKC.PDO is used to partially decrypt the ciphertext $[x']_{PK_1}$, $[y']_{PK_1}, [S]_{PK_1}, [T]_{PK_2}$ with λ_1 to get the results $[x']^1_{PK_1}, [y']^1_{PK_2}, [S]^1_{PK_1}$, $[T]^1_{PK_2}$. Finally, CSS packs the ciphertext $[x']_{PK_1}, [y']_{PK_1}, [S]_{PK_1}$, $[T]_{PK_2}$
  and the partially decrypted ciphertext $[x']^1_{PK_1}, [y']^1_{PK_2}, [S]^1_{PK_1}, [T]^1_{PK_2}$ as $([x']_{PK_1}, [y']_{PK_1}, [S]_{PK_1}, [T]_{PK_2}, [x']^1_{PK_1}, [y']^1_{PK_2}, [S]^1_{PK_1}, [T]^1_{PK_2})$ and sends them to CSP.
- Upon receiving $([x']_{PK_1}, [y']_{PK_2}, [S]_{PK_1}, [T]_{PK_2}, [x']^1_{PK_1}, [y']^1_{PK_2}, [S]^1_{PK_1}$, $[T]^1_{PK_2})$, CSP furtherly decrypts the ciphertext $[x']^1_{PK_1}, [y']^1_{PK_2}$, $[S]^1_{PK_1}, [T]^1_{PK_2}$ with λ_2 by DT-PKC.PDT to get the plaintext x', y', S, T. Then, it computes $h = x' \cdot y'$, and uses the public PK_3 to encrypt h, S, T by DT-PKC.Enc respectively. Finally, CSP sends $[h]_{PK_3}, [S]_{PK_3}, [T]_{PK_3}$ to CSS.
- Upon receiving the ciphertext $[h]_{PK_3}, [S]_{PK_3}, [T]_{PK_3}$, CSS removes the random number r_x, r_y, R_x, R_y by computing $S_3 = [R_x]^{N-1}_{PK_1}, S_4 = [R_y]^{N-1}_{PK_2}, S_5 = [r_x \cdot r_y]^{N-1}$. Finally, CSS computes and outputs the ciphertext $[z] = H \cdot S_1 \cdot S_2 \cdot S_3 \cdot S_4 \cdot S_5$ as the product.

The basic protocol works as Algorithm 2.

3.4 Secure Vector Inner Product (SVIP)

The goal of SVIP is to input two encrypted vectors $[X]_{PK_1} = ([x_1], \ldots, [x_n])$, $[Y]_{PK_2} = ([y_1], \ldots, [y_n])$ and compute the ciphertext $[Z]_{PK_3} = [\sum_{i=1}^{n}(x_i \cdot y_i)]_{PK_3}$ without revealing the plaintext. The basic protocol is described in Algorithm 3.

3.5 Secure Sigmoid Function (Ssigm)

The goal of Ssigm is to input two vectors $[x_i]_{PK_1}$, $[w_i]_{PK_2}$ and compute $[Z]_{PK_3} = [\frac{1}{1+e^{-(\Sigma_{i=1}^{n} x_i \cdot w_i)}}]_{PK_3}$. Since the sigmoid function cannot be directly supported by HE, we approximate it using Taylor expansion. The calculation process is completed by CSS and CSP as follows:

Algorithm 2. Secure Integer Multiplication

Input: The ciphertexts $[x]_{PK_1}$ and $[y]_{PK_2}$, and
public keys PK_1, PK_2, PK_3

Output: $[z]_{PK_3} = [x \cdot y]_{PK_3}$

@CSS

1. Select random numbers $r_x, r_y, R_x, R_y \in Z_N$ and encrypt them
2. Calculate: $[x']_{PK_1} = [x]_{PK_1} \cdot [r_x]_{PK_1}$;
$\qquad [y']_{PK_2} = [y]_{PK_2} \cdot [r_y]_{PK_2}$;
$\qquad [S]_{PK_1} = [R_x]_{PK_1} \cdot [x]_{PK_1}^{N-r_y}$;
$\qquad [T]_{PK_2} = [R_y]_{PK_2} \cdot [y]_{PK_2}^{N-r_x}$
3. Partitial decryption: $[x']_{PK_1}^1 = DT - PKC.PDO([x']_{PK_1})$;
$\qquad [y']_{PK_2}^1 = DT - PKC.PDO([y']_{PK_2})$;
$\qquad [S]_{PK_1}^1 = DT - PKC.PDO([S]_{PK_1})$;
$\qquad [T]_{PK_2}^1 = DT - PKC.PDO([T]_{PK_2})$
4. Send $([x']_{PK_1}, [y']_{PK_2}, [S]_{PK_1}, [T]_{PK_2}, [x']_{PK_1}^1, [y']_{PK_2}^1,$
$\qquad [S]_{PK_1}^1, [T]_{PK_2}^1)$ to CSP

@CSP

1. Calculate $x' = DT - PKC.PDT([x']_{PK_1}, [x']_{PK_1}^1)$;
$\qquad y' = DT - PKC.PDT([y']_{PK_2}, [y']_{PK_2}^1)$;
$\qquad S = DT - PKC.PDT([S]_{PK_1}, [S]_{PK_1}^1)$;
$\qquad T = DT - PKC.PDT([T]_{PK_2}, [T]_{PK_2}^1)$
2. Calculate $h = x' \cdot y'$
3. Encrypt to get $H = DT - PKC.Enc(h)$;
$\qquad S_1 = DT - PKC.Enc(S)$;
$\qquad S_2 = DT - PKC.Enc(T)$
3. Send H, S_1, S_2 to CSS

@CSS

1. Calculate $S_3 = [R_x]^{N-1}, S_4 = [R_y]^{N-1}, S_5 = [r_x \cdot r_y]^{N-1}$
2. Return $[z]_{PK_3} = H \cdot S_1 \cdot S_2 \cdot S_3 \cdot S_4 \cdot S_5$

- CSS computes $[x_i \cdot w_i]$ by Algorithm 3 to obtain the ciphertext $[v] = [\Sigma_{i=1}^n(x_i \cdot w_i)]$. Next, it calls DT-PKC.PDO to partially decrypt the ciphertext $[v]$ with its partial master key λ_1 and gets the results $[v]_1$. Finally, CSS sends the message $([v], [v]_1)$ to CSP.
- Upon receiving the message $([v], [v]_1)$, CSP furtherly decrypts the ciphertext $[v]$ with λ_2 by calling algorithm DT-PKC.PDT. Then, it computes $e^{-v} \approx 1 - v + \frac{1}{2}v^2 - \frac{1}{6}v^3 + \frac{1}{24}v^4 - \frac{1}{120}v^5$ and obtain $Z = \frac{1}{1+e^{-v}} = \frac{1}{2} + \frac{1}{4}v - \frac{1}{48}v^3 + \frac{1}{480}v^5 - O(v^7) \approx \frac{1}{2} + \frac{1}{4}v - \frac{1}{48}v^3 + \frac{1}{480}v^5$. Finally, CSP sends the message Z to CSS.
- Upon receiving Z, CSS encrypts Z using PK_3 to get $[Z]_{PK_3}$ and outputs it. The protocol is described in Algorithm 4.

Algorithm 3. Secure Vector Inner Product

Input: The ciphertexts $[X]_{PK_1}$ and $[Y]_{PK_2}$, and
public keys PK_1, PK_2, PK_3

Output: $[Z]_{PK_3} = [\sum_{i=1}^{n}(x_i \cdot y_i)]_{PK_3}$

@CSS

  1. Select random numbers $r_x^i, r_y^i, R_x^i, R_y^i \in Z_N (i \in \{1, \ldots, n\})$
    and encrypt them

  2. For $i{=}1$ to n do

  3. Calculate: $[x_i']_{PK_1} = [x_i]_{PK_2} \cdot [r_x^i]_{PK_1}$;
        $[y_i']_{PK_2} = [y_i]_{PK_2} \cdot [r_y^i]_{PK_2}$;
        $[S_i]_{PK_1} = [R_x^i]_{PK_1} \cdot [x_i]_{PK_1}^{N-r_y^i}$;
        $[T_i]_{PK_2} = [R_y^i]_{PK_2} \cdot [y_i]_{PK_2}^{N-r_x^i}$

  4. Partitial decryption: $[x_i']_{PK_1}^1 = DT - PKC.PDO([x_i']_{PK_2})$;
        $[y_i']_{PK_2}^1 = DT - PKC.PDO([y_i']_{PK_2})$;
        $[S_i]_{PK_1}^1 = DT - PKC.PDO([S_i]_{PK_1})$;
        $[T_i]_{PK_2}^1 = DT - PKC.PDO([T_i]_{PK_2})$

  5. End for

  6. Send $([x_i']_{PK_2}, [y_i']_{PK_2}, [S_i]_{PK_1}, [T_i]_{PK_2}, [x_i']_{PK_1}^1, [y_i']_{PK_2}^1,$
    $[S_i]_{PK_1}^1, [T_i]_{PK_2}^1)$ to CSP

@CSP

  1. For $i{=}1$ to n do

  2. Calculate $x_i' = DT - PKC.PDT([x_i']_{PK_1}, [x_i']_{PK_1}^1)$;
        $y_i' = DT - PKC.PDT([y_i']_{PK_2}, [y_i']_{PK_2}^1)$;
        $S_i = DT - PKC.PDT([S_i]_{PK_1}, [S_i]_{PK_1}^1)$;
        $T_i = DT - PKC.PDT([T_i]_{PK_2}, [T_i]_{PK_2}^1)$

  3. Calculate $h_i = x_i' \cdot y_i'$

  4. Encrypt $[h_i]_{PK_3} = DT - PKC.Enc(h_i)$;
        $[S_i]_{PK_3} = DT - PKC.Enc(S_i)$;
        $[T_i]_{PK_3} = DT - PKC.Enc(T_i)$

  5. End for

  6. Send $[h_i]_{PK_3}, [S_i]_{PK_3}, [T_i]_{PK_3}$ to CSS

@CSS

  1. For $i{=}1$ to n do

  2. Calculate $S_3^i = [R_x^i]_{PK_1}^{N-1}, S_4^i = [R_y^i]_{PK_2}^{N-1}$,
        $S_5^i = [r_x^i \cdot r_y^i]_{PK_1}^{N-1}$

  3. Calculate $[Z]_{PK_3} = [h_i]_{PK_3} \cdot [S_i]_{PK_3} \cdot [T_i]_{PK_3} \cdot S_3^i \cdot S_4^i \cdot S_5^i$

  4. Return $[Z]_{PK_3}$

Algorithm 4. Secure Sigmoid

Input: The ciphertexts $[x_i]_{PK_1}$, $[w_i]_{PK_2}$, and
       public keys PK_1, PK_2, PK_3

Output: $[Z]_{PK_3}$

@CSS
  1. For $i=1$ to n do
  2. $[v]_{PK_3}=\boldsymbol{SVIP}([x_i]_{PK_1},[w_i]_{PK_2})$
  3. $[v]^1_{PK_3} = DT-PKC.PDO([v]_{PK_3})$
  4. End for
  5. Send $[v]_{PK_3},[v]^1_{PK_3}$ to CSP

@CSP
  1. $v= DT-PKC.PDT([v]_{PK_3},[v]^1_{PK_3})$
  2. $e^{-v} = 1 - v + 1/2v^2 - 1/6v^3 + 1/24v^4 - 1/120v^5$;
  3. $Z = 1/(1+e^{-v})$
  4. Send Z to CSS;

@CSS
  1. Encrypt to get $[Z]_{PK_3}$
  2. Return $[Z]_{PK_3}$

3.6 Privacy-Preserving LR Training

To enhance the efficiency of training, some parameters are published before encrypting the data, such as floating-point number precision E, learning rate α, and maximum number of iterations T. In order to train the model, the training entity CSS needs to run the RGA and parameter updating algorithms with the assistance of CSP. The concrete algorithm of privacy-preserving logistic regression (PPLR) is shown in Algorithm 5.

3.7 Discussion

Without affecting the communication efficiency, the security level increases with the number of chosen random numbers. In our implementation, we choose four random numbers, allowing the protocol to maintain the same number of communication rounds as the multiplication protocol in [16], but our scheme is more secure.

Algorithm 5. EPoLORE

Input: $A = \{([x_{11}], ..., [x_{1d}], [y_1]), , ([x_{s1}], ..., [x_{sd}], [y_s])\}, \alpha, T, E$
Output: $[w_j]$
Initialize: $[w_j] = 0$
  1. for $j = 1$ to T do
  2.   for $i=1$ to s do
  3. $\alpha = 4/(1.0 + j + i) + 0.01$
  4. $h = \textbf{\textit{Ssigm}}([w_i], [x_i], PK_1, PK_2, PK_3)$
  5.   if $h > 0.5$
  6. $p_i = 0$
  7.   else $p_i = 1$
  8.   end if
  9. $[p_i] = DT - PKC.Enc(p_i)$
 10. $[error] = [p_i] \cdot [y_i]^{N-1} \mod N^2$
 11. $q = [error]^\alpha$
 12. $[\theta] = \textbf{\textit{SIM}}(q, [x_{ij}])$
 13. $\theta = DT - PKC.Dec([\theta])$
 14.   if $\theta \neq 0$
 15. $[w_j] = [w_j] \cdot [\theta] \mod N^2$
 16.   end if
 17. end for
 18. end for
 19. Return $[w_j]$

---

4   Security Analysis

In this section, we will analyze the security of our proposed arithmetic protocols and the EPoLORE.

4.1   Security of Arithmetic Protocols

We now prove the security of our proposed arithmetic protocols. Given the specific context of the system, correctness can be drawn from the design of the SIM, SVIP, and Ssigm protocols presented in this paper. For privacy of the protocols, we have the following theorems.

Theorem 1. *Assuming that DT-PKC is semantically secure [16], the SIM can securely compute multiplication of ciphertexts in the presence of semi-honest adversaries under the real/ideal paradigm.*

Proof. We prove the security in case of CSS and CSP corrupted, respectively.
   Suppose $\mathcal{A}$ controls CSS. We construct a simulator $\mathcal{S}_1$ as follows:

1. $\mathcal{S}_1$ initializes all parameters.
2. $\mathcal{S}_1$ invokes $\mathcal{A}$ to get $([x']_{PK_1}, [y']_{PK_1}, [S]_{PK_1}, [T]_{PK_2}, [x']^1_{PK_1}, [y']^1_{PK_2}, [S]^1_{PK_1}, [T]^1_{PK_2})$;

3. $\mathcal{S}_1$ checks all the messages received, and will send $\perp$ to the trusted party if the check does not pass. At this point, an interrupt is simulated, the output of $\mathcal{A}$ is output;
4. Otherwise, $\mathcal{S}_1$ computes $[x']^1_{PK_1}, [y']^1_{PK_2}, [S]^1_{PK_1}, [T]^1_{PK_2}$ successively and performs the $DT - PKC.PDO$ to get h, S, T.
5. $\mathcal{S}_1$ encrypts h, S, T using PK_3, respectively, then get $[h]_{PK_3}, [S']_{PK_3}, [T']_{PK_3}$.
6. $\mathcal{S}_1$ outputs whatever $\mathcal{A}$ outputs.

The CSS's view in the real-world execution of the protocol is $([h]_{PK_3}, [S]_{PK_3}, [T]_{PK_3}, [x \cdot y]_{PK_3})$. Based on the aforementioned simulation, it is obvious that the view of $\mathcal{A}$ is same, except the construction of S'_1, S'_2. Since S', T' and S, T in the scheme are all random numbers belonging to the group, and S, T, S', T' are all ciphertexts in the protocol, S' is indistinguishable from S, either T' from T.

When $\mathcal{A}$ controls the CSP, the simulator $\mathcal{S}_2$ works as follows:

1. $\mathcal{S}_2$ computes $\overline{[x']}, \overline{[y']}, \overline{[S]}$ and $\overline{[T]}$ successively, and then runs as CSS to get $([x']_1, [y']_1, [S]_1, [T]_1)$, and sends $([x']_1, [y']_1, [S]_1, [T]_1)$ to $\mathcal{A}$.
2. $\mathcal{S}_2$ receives $[h]_{PK_3}, [S']_{PK_3}, [T']_{PK_3}$ from $\mathcal{A}$.
3. $\mathcal{S}_2$ outputs whatever $\mathcal{A}$ outputs.

CSP's view in the real-world execution of the protocol is $([x'], [y'], [S], [T], [x']_1, [y']_1, [S]_1, [T]_1)$. From above simulation, $([x'], [y'], [S], [T])$ is indistinguishable from $(\overline{[x']}, \overline{[y']}, \overline{[S]}, \overline{[T]})$ due to the semantic security of DT-PKC. Thus the view of the adversary $\mathcal{A}$ in the real-word execution of the protocol is indistinguishable from that of the ideal-word execution of the protocol.

Theorem 2. *Assuming DT-PKC is semantically secure, the SVIP protocol can securely perform vector inner product operations in the presence of semi-honest adversaries.*

Proof. The proof of this theorem is similar to the proof of Theorem 1 and we omit it here.

4.2 Security of EPoLORE

In EPoLORE, we assume the adversary can control the communication channels between the SF and the CSS. The adversary can also intercept all the data transmitted, but is unable to decrypt it due to their lack of knowledge of the SF's private key. Adversary is able to obtain the data transmitted between the CSS/SP and the CSP/CSS, but he cannot know the private key of the challenging party (CSS, CSP and SP). The security and privacy of train and classification are guaranteed by the semantic security of DT-PKC.

The adversary may appear during the computation of the vector inner product, activation function, calculation of θ values, and the final update of model parameters. Since all the data involved in these computations are ciphertexts, the adversary cannot access the original data. Of course, the adversary may attempt to corrupt the CSS or CSP to obtain λ_1 or λ_2, but he still cannot recover any plaintext since he does not know the other part of the decryption key. The decrypted data has been blinded through random numbers, making it difficult for the adversary to obtain the original plaintext. Therefore, the EPoLORE proposed in this paper can preserve the privacy of data during the model training.

5 Performance Evaluation

In this section, we will evaluate the performance of the proposed EPoLORE.

5.1 Implementation

The experiments were conducted on a PC with system windows 10, processor $Intel(R)Core(TM)i7 - 1165G7@2.80GHz1.69GHz$ and $16.0GB$ of RAM on board. We selected two datasets, namely IRIS and HEART disease, for our experiments.

5.2 Experimental Results

We give the classification results and accuracy rates for training. Firstly, we show the classification of EPoLORE on dataset IRIS and HEART, as Fig. 2. It can be seen the three categories are separated out clearly, so does the Heart dataset.

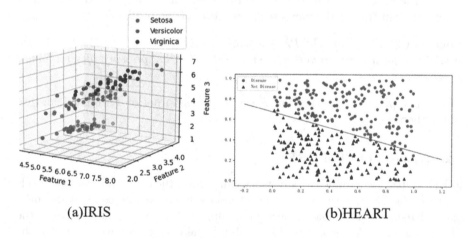

(a)IRIS   (b)HEART

Fig. 2. Classification of dataset.

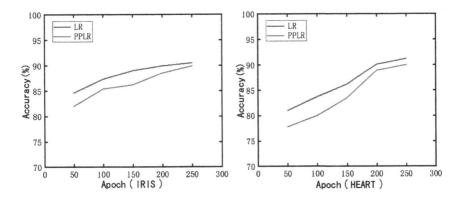

Fig. 3. Accuracy of EPoLORE.

We evaluated the trend of accuracy with the number of iterations on the Iris dataset. The evaluation results of the accuracy are shown in Fig. 3.

It can be seen from Fig. 3 that the classification accuracy of the PPLR and LR are comparable for the same number of iterations. The reason for the decline in the PPLR of classification accuracy is mainly for Taylor approximation, but the decrease is not significant.

Besides accuracy, we also considered the metrics such as the recall, precision, and F1 score. The results are shown in Fig. 4. It can be seen that in the EPoLORE, these metrics slightly decrease compared to the plaintext training on dataset HEART. The results indicate that the model is feasible in classifying with privacy-preserving dataset.

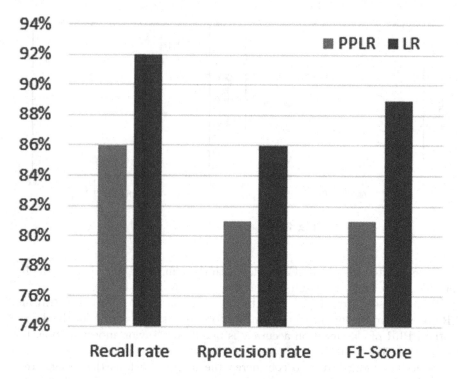

Fig. 4. Evaluation metrics.

6   Conclusion

In this paper, we proposed EPoLORE, a novel privacy-preserving logistic regression scheme in cloud environment. In order to protect the privacy of data, while achieving similar levels of accuracy as unencrypted logistic regression models, we designed efficient interactive protocols for integer and floating-point computations, including integer multiplication, vector inner product, and the activation function calculation. Based on the interactive protocols mentioned above, the Logistic Regression in ciphertext can be implemented with the combination and iterations of the calculations. The security of the protocols is analyzed and thereby the security of the EPoLORE. We evaluate the performance of our protocol using two real datasets, and the experiments show the scheme achieves privacy-preserving of the training data, while maintaining the reasonable accuracy of the model.

References

1. Jain, D., Singh, V.: A two-phase hybrid approach using feature selection and adaptive SVM for chronic disease classification. Int. J. Comput. Appl. **43**, 524–536 (2021)

2. Dumitrescu, E., Hué, S., Hurlin, C., Tokpavi, S.: Machine learning for credit scoring: improving logistic regression with non-linear decision-tree effects. Eur. J. Oper. Res. **297**, 1178–1192 (2022)
3. Szafraniec-Siluta, E., Zawadzka, D., Strzelecka, A.: Application of the logistic regression model to assess the likelihood of making tangible investments by agricultural enterprises. Procedia Comput. Sci. **207**, 3894–3903 (2022)
4. Han, K., Hong, S., Cheon, J.H., Park, D.: Logistic regression on homomorphic encrypted data at scale. In: Proceedings of the AAAI Conference on Artificial Intelligence, vol. 33, no. 01, 9pp. 466–9471 (2019)
5. Mandal, K., & Gong, G.: PrivFL: practical privacy-preserving federated regressions on high-dimensional data over mobile networks. In: Proceedings of the 2019 ACM SIGSAC Conference on Cloud Computing Security Workshop, pp. 57–68 (2019)
6. Mohassel, P., Zhang, Y.: SecureML: a system for scalable privacy-preserving machine learning. In: 2017 IEEE Symposium on Security and Privacy, pp. 19–38 (2017)
7. Cheon, J.H., Kim, D., Kim, Y., Song, Y.: Ensemble method for privacy-preserving logistic regression based on homomorphic encryption. IEEE Access **6**, 46938–46948 (2018)
8. Shi, H., et al.: Secure multi-pArty computation grid LOgistic REgression (SMAC-GLORE). BMC Med. Inform. Decis. Mak. **16**, 175–187 (2016)
9. Li, J., et al.: Efficient and secure outsourcing of differentially private data publishing with multiple evaluators. IEEE Trans. Dependable Secure Comput. **19**(1), 67–76 (2020)
10. Zhu, L., Tang, X., Shen, M., Gao, F., Zhang, J., Du, X.: Privacy-preserving machine learning training in IoT aggregation scenarios. IEEE Internet Things J. **8**(15), 12106–12118 (2021)
11. Aono, Y., Hayashi, T., Trieu Phong, L., Wang, L.: Scalable and secure logistic regression via homomorphic encryption. In: Proceedings of the Sixth ACM Conference on Data and Application Security and Privacy, pp. 142–144 (2016)
12. Carpov, S., Gama, N., Georgieva, M., Troncoso-Pastoriza, J.R.: Privacy-preserving semi-parallel logistic regression training with fully homomorphic encryption. Cryptology ePrint Archive (2019)
13. Cheon, J.H., Kim, A., Kim, M., Song, Y.: Homomorphic encryption for arithmetic of approximate numbers. In: Takagi, T., Peyrin, T. (eds.) ASIACRYPT 2017. LNCS, vol. 10624, pp. 409–437. Springer, Cham (2017). https://doi.org/10.1007/978-3-319-70694-8_15
14. Liu, X., Deng, R.H., Choo, K.K.R., Weng, J.: An efficient privacy-preserving outsourced calculation toolkit with multiple keys. IEEE Trans. Inf. Forensics Secur. **11**, 2401–2414 (2016)
15. Sebbouh, O., Cuturi, M., Peyré, G.: Randomized stochastic gradient descent ascent. In: International Conference on Artificial Intelligence and Statistics, pp. 2941–2969 (2022)
16. Goldreich, O.: Foundations of Cryptography: Volume 2, Basic Applications. Cambridge University Press, Cambridge (2009)
17. Wang, J., Wu, L., Wang, H., Choo, K.K.R., He, D.: An efficient and privacy-preserving outsourced support vector machine training for internet of medical things. IEEE Internet Things J. **8**(1), 458–473 (2020)

Multi-dimensional Data Aggregation Scheme without a Trusted Third Party in Smart Grid

Sucheng Yan⬤, Yuhong Sun(✉) ⬤, Wendan Zhang, Yining Liu, Chen Zhang, and Hua Wang

School of Computer Science, Qufu Normal University, Rizhao, China
sun_yuh@163.com

Abstract. The data aggregation is an essential requirement in the smart grid that is gradually replacing the traditional grid for its reliability, flexibility, efficiency and other excellent performance. The one-dimensional data aggregation schemes cannot meet the requirements of fine-grained analysis, while most multi-dimensional data aggregation schemes rely on a trusted third party (TTP). Besides, some multi-dimensional data aggregation schemes cannot resist the collusion attacks from some users, the aggregator and the control center. In order to satisfy the fine grained analysis of the electricity consumption data and resist the collusion attack, meanwhile alleviate the dependence on the TTP, we propose a multi-dimensional data aggregation scheme based on the EC-ElGamal cryptosystem and the super increasing sequences. The scheme can achieve fine-grained data analysis of multi-dimensional electricity data, and each entity generates its own keys without any TTP, which avoids users' privacy leakage. Furthermore, our scheme can resist the replay attack and the collusion attack from some users, the control center and the aggregator. Security analysis shows that the proposed scheme can meet the security requirements of the smart grid and is more practical than the existing schemes.

Keywords: Smart grid · Privacy-preserving · Multi-dimensional data aggregation · No trusted third party

1 Introduction

Compared with the traditional power grid, the smart grid is superior in reliability, flexibility and efficiency. In the smart grid, a large number of smart meters are deployed in the user residential areas, industrial areas, and commercial markets, through which the electricity consumption data of users is measured and reported periodically. Automatically collecting the user electricity data is a basic requirement in the smart grid [1]. With the help of specific and refined user data, electricity companies can adopt optimal strategies to control power production and distribution, to analyze power data and adjust electricity price dynamically. However, direct delivery of data to electricity companies implies processing a large amount of user electricity data in a short period of time, and will cause great pressure for the communication. Moreover, the direct submission of

© The Author(s), under exclusive license to Springer Nature Singapore Pte Ltd. 2024
J. Vaidya et al. (Eds.): AIS&P 2023, LNCS 14510, pp. 78–91, 2024.
https://doi.org/10.1007/978-981-99-9788-6_7

electricity consumption data will also leak the user's privacy. Users' electricity consumption data usually contains private information, such as users' habits and lifestyles, which may cause serious consequences if a malicious attacker obtains such information [2]. Therefore, how to balance the availability and privacy of electricity consumption data is not only an important academic issue, but also a technical bottleneck in the smart grid. In view of the above problems, privacy-preserving data aggregation is considered as a feasible solution.

Most of the traditional privacy-preserving data aggregation schemes in the smart grid aim to aggregate one-dimensional data [3–5], while the one-dimensional data aggregation cannot satisfy the fine-grained analysis of the control center. However, some existing multi-dimensional data aggregation schemes [8, 11, 13, 18] rely on a TTP. If the TTP is attacked, the adversary can obtain the user's data easily. At the same time, the existing multi-dimensional data aggregation schemes [12, 14, 18] cannot resist the collusion attack from some users, the control center and the aggregator. To solve the above problems, we propose a privacy-preserving multi-dimensional data aggregation scheme without a TTP in the smart grid.

1.1 Related Work

There have been a lot of research on the secure data aggregation in the smart grid, such as schemes [6–13] based on homomorphic encryption and schemes [14–18] based on blind factors.

Homomorphic encryption is by far the most popular method of data aggregation for privacy protection. Lu et al. [6] encrypted the multi-dimensional data in the local gateway, and the control center obtained the aggregated electricity consumption data. Compared with the traditional one-dimensional aggregation schemes, the scheme has the lower computation cost and the higher communication efficiency. Later, they achieved a two-dimensional data aggregation scheme [7]. However, the scheme [7] cannot meet the requirements of data integrity and authentication. Boudia et al. [8] proposed a multi-dimensional data aggregation scheme based on elliptic curve cryptography (ECC). The scheme does not require bilinear pairing operations and has a low computation cost. However, the scheme relies on a TTP. Shen et al. [9] also proposed a data aggregation scheme based on Paillier's homomorphic encryption. This scheme uses Horner's rule [10] to compress the multi-dimensional data structure together, and designs an aggregation scheme that supports multiple types of data. Chen et al. [11] proposed another data aggregation scheme based on the Paillier's homomorphic encryption. Their scheme enables smart meters to report multiple types of data in a single message, and electricity companies can analyze the variance on the data. Similar to [8], the shortcoming is that the scheme has to rely on a TTP. Zeng et al. [12] proposed a multi-subset data aggregation scheme based on Paillier's homomorphic encryption without a TTP, and the control center can obtain the total electricity consumption of each subset. However, the scheme does not consider the integrity of the data, and it cannot resist the collusion attack launched by users, aggregators and control centers on a single user. Zuo et al. [13] proposed a multi-dimensional data aggregation scheme without a TTP, while the registration phase of the scheme cannot resist the replay attack.

Some schemes enable data aggregation in smart grids using blind factors. The TTP generates a series of random numbers (called blind factors) and securely distributes them to users and aggregators. The user confuses the electricity data with his own blind factor, and the aggregator eliminates all user-added blind factors to obtain the correct aggregated data. Fan et al. [14] proposed the first data aggregation scheme based on the blind factors that can resist internal attacks. Bao et al. [15] found the problem of key leakage and data integrity in [14]. With the BGN cryptosystem and blind factors, He et al. [16] can guarantee the integrity of data, and the computation cost is much lower than that proposed by [14]. While the scheme proposed by Bao et al. [17] is fault-tolerant, and the aggregated data can be recovered when smart meters break down in the system. Based on ECC, Ming et al. [18] designed a privacy-preserving data aggregation scheme that supports multiple types of data transmission. It enables the control center to perform fine-grained analysis of the electricity data. But the scheme must rely on a TTP to generate the blind factors.

1.2 Contributions

The main contributions of this paper can be summarized as follows.

1) We propose a multi-dimensional data aggregation scheme without a TTP in the smart grid. The keys of each entity are generated just by himself, which reduces the dependence on the TTP.
2) We consider the collusion attack launched from users, the fog node and the control center, in which case colluders also cannot obtain the private data of a single user. No one can obtain the encrypted data of the user except himself. And our scheme can resist the replay attack.
3) We analyze the security from the privacy, the integrity and the authentication. The experiments show the computation and communication costs are relatively low.

2   Preliminaries and System

2.1   EC-ElGamal [19] Cryptosystem

KeyGen: Given a group of order q on an elliptic curve, the generator of the group is P, randomly choose a private key $x \in Z_q^*$, calculate the public key $Y = x \cdot P$.

Encrypt: Given a short message m, randomly choose a number r, calculate the ciphertext as $C = (C_1, C_2) = (r \cdot P, m \cdot P + r \cdot Y)$.

Decrypt: Given a ciphertext $C = (C_1, C_2)$, compute $m \cdot P = Dec(C_1, C_2) = (C_2 - xC_1)$, m is recovered with Pollard's lambda algorithm [20].

Homomorphic encryption: For m_1 and m_2, the EC-ElGamal cryptosystem has the homomorphism property:

$$Enc(m_1 + m_2) = Enc(m_1) + Enc(m_2) = (C_{1,a} + C_{2,a}, C_{1,b} + C_{2,b})$$
$$= ((r_1 + r_2) \cdot P, (m_1 + m_2) \cdot P + (r_1 + r_2) \cdot Y)$$

2.2 BLS [21] Short Signature

Initialize: Given two cyclic groups G_1, G_2 of the prime order p based on an elliptic curve, P is the generator of the cyclic group G_1, $e : G_1 \times G_1 \rightarrow G_2$ is a bilinear pair, and $H : \{0, 1\}^* \rightarrow G_1$ is a hash function.

KeyGen: Randomly choose a private key $x \in Z_p^*$, and compute the public key as $Y = xP$.

Sign: Given a message m, first map it to a point on the elliptic curve by $H:h = H(m)$, and output the signature $\sigma = xh$.

Verify: Check the equation $e(P, \sigma) = e(Y, h)$. If it holds, output "accepted".

2.3 System Model

As shown in Fig. 1, the system model of the proposed scheme involves three types of entities: smart meter (SM), fog node (FN), and control center (CC). For the sake of clarity, we assume that a number of geographically close users form a group, such as a residential area, and each user has one SM. A FN is responsible for data aggregation and transmission in such a residential area. The entities in a smart grid system are described as follows.

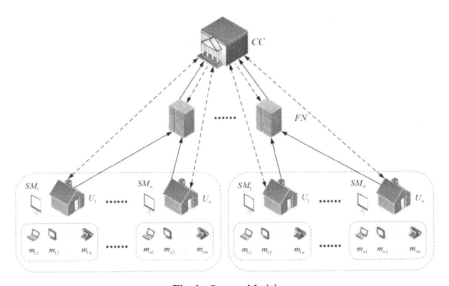

Fig. 1. System Model.

Each SM collects and aggregates the multiple types of the electricity consumption data from a user, such as air conditioning electricity data, etc. SM then uploads them to FN within a short time.

FN performs the verification and aggregation of data from SMs, preprocesses and aggregates them before sending them to the control center.

CC maintains the status of the entire smart grid, which has sufficient computing and storage resources to decrypt aggregated data, analyze the data, adjust pricing strategies, and generate parameters for the entire system.

Among the entities, CC is considered to be honest but curious. FN can be easily controlled by adversaries, and is considered untrustworthy. SMs are often seen as honest but curious. In addition, we consider the worst-case scenario in which CC, FN, and some SMs collude to obtain the electricity consumption data of other SMs.

3   The Proposed Scheme

The proposed scheme includes 6 stages: system initialization, authentication, group public key generation, data encryption, data aggregation, and data decryption.

3.1   Initialization

In this stage, CC generates the system parameters, including two elliptic curve cyclic groups G, G_T. Assume there are w dimensions of the electricity consumption data to be analyzed. The other system parameters are generated as the following steps. Given an elliptic curve cyclic group G of order q(a large prime), with the generator P.

1) Chooses a random number $x_{CC} \in Z_q^*$ as its private key, and calculates its public key $pk_{CC} = x_{CC} \cdot P$.
2) Chooses a bilinear mapping $e : G \times G \to G_T$, and a hash function: $H : \{0, 1\}^* \to G$.
3) Chooses k consecutive range $[R_1, R_2), \ldots, [R_2, R_3), [R_k, T)$, where $R_1 = 0, T << q$, T is assumed to be the upper bound of electricity consumption. Then, generates two super increasing sequences, of which the first sequence is $\overrightarrow{a} = \{a_1, a_2, \ldots, a_w\}$, where $a_1, a_2, \ldots, a_w$ are prime numbers, satisfying $\sum_{j'=1}^{j-1} a_{j'} \cdot n \cdot T < a_j, j = 1, 2, \ldots, w, \sum_{j=1}^{w} a_j \cdot n \cdot T << q$. The second sequence is $\overrightarrow{b} = \{b_1, b_2, \ldots, b_k\}$, satisfying $q >> b_u > b_0 + b_{u-1} \cdot n, b_0 > \sum_{j=1}^{w} a_j \cdot n \cdot T, u = 1, 2, \ldots, k$.
4) Publishes the parameters $\{G, G_T, P, e, H, pk_{CC}, \overrightarrow{a}, \overrightarrow{b}, (R_1, R_2, \ldots, R_k, T)\}$.

3.2   Authentication

1) SM_i chooses a random number $x_i \in Z_q^*$ as its private key, computes the public key as $pk_i = x_i \cdot P$. Then, SM_i computes the signature $\sigma_i = x_i \cdot H(ID_i || T_i)$, where ID_i is the user's identity and T_i is the timestamp. After that, SM_i sends $(ID_i, T_i, \sigma_i, pk_i)$ to CC.
2) CC checks the validity of T_i. If it is valid, CC verifies whether $e(P, \sigma_i) = e(pk_i, H(ID_i || T_i))$ holds. And if so, CC issues the certificate $Cert_i = x_{CC} \cdot H(ID_i || pk_i || T_i)$ to the SM_i, where x_{CC} is the private key of CC.
3) The authentication of FN is similar to that of SM_i.

3.3 Group Public Key Generation

In this process, all SMs broadcast their public keys in the group firstly. It is assumed that there is at least one honest SM. Then, the SMs jointly generate the group public key PK. The specific process is as follows:

1) SM_i broadcasts ID_i, pk_i, $Cert_i$ in the group.
2) SM_i verifies the identities of other group members with the following equation:

$$\prod_{\substack{j=1 \\ j \neq i}}^{n} e(H(ID_j||pk_j||T_j), pk_{CC}) = e(P, \sum_{\substack{j=1 \\ j \neq i}}^{n} Cert_j) \tag{1}$$

3) If SM_i can verify the identities of other group members, it calculates the group public key PK by the following equation:

$$PK = \sum_{i=1}^{n} pk_i = (x_1 + x_2 + \cdots + x_n) \cdot P \tag{2}$$

3.4 Data Encryption

According to the parameters published by CC, SM_i determines which range its total electricity consumption data belongs to: suppose the electricity consumption data of a user U_i is $(m_{i1}, m_{i2}, \ldots, m_{iw})$, and the total electricity consumption data is $m_i = m_{i1} + m_{i2} + \cdots + m_{iw}$; If m_i satisfying $R_u \leq m_i < R_{u+1}$, m_i is in the range $[R_u, R_{u+1})$, and S_{N_u} is the number of users with m_i in the range $[R_u, R_{u+1})$, $u = 1, 2, \ldots, k$.

1) Suppose m_i is a short message, if $R_u \leq m_i < R_{u+1}$, SM_i chooses a random number $r_i \in Z_q^*$, using PK and b_u to compute the ciphertext, where b_u is the parameter of the corresponding range of the total electricity consumption data of SM_i from $\overrightarrow{b}$, $u = 1, 2, \ldots, k$. The ciphertext is $(C_{i,a}, C_{i,b})$:

$$\begin{aligned} C_{i,a} &= r_i \cdot P \\ C_{i,b} &= (a_1 m_{i1} + a_2 m_{i2} + \cdots a_w m_{iw}) \cdot P + b_u \cdot P + r_i \cdot PK \end{aligned} \tag{3}$$

2) SM_i uses to compute the signature:

$$\sigma_i' = x_i H(ID_i||C_{i,a}||C_{i,b}||T_i') \tag{4}$$

3) SM_i packs and sends $\{ID_i, C_{i,a}, C_{i,b}, T_i', \sigma_i'\}$ to FN.

3.5 Data Aggregation

After receiving $\{ID_i, C_{i,a}, C_{i,b}, T_i', \sigma_i'\}$ from SM_i, FN performs the following:

1) FN checks the validity of the timestamp T_i'. If it is valid, FN performs batch verification operation on signatures of n SMs:

$$e(P, \sum_{i=1}^{n} \sigma_i') = \prod_{i=1}^{n} e(H(ID_i||C_{i,a}||C_{i,b}||T_i'), pk_i) \tag{5}$$

2) If the signature is verified, FN computes the aggregated ciphertext (C_a, C_b) as follows.

$$(C_a, C_b) = (\sum_{i=1}^{n} C_{i,a}, \sum_{i=1}^{n} C_{i,b})$$

$$= (\sum_{i=1}^{n} r_i \cdot P, (a_1 \sum_{i=1}^{n} m_{i1} + a_2 \sum_{i=1}^{n} m_{i2} + \cdots + a_w \sum_{i=1}^{n} m_{iw}) \cdot P \tag{6}$$

$$+ (b_1 S_{N_1} + b_2 S_{N_2} + \cdots + b_k S_{N_k}) \cdot P + \sum_{i=1}^{n} r_i \cdot PK)$$

3) Signs on (6) as $\sigma'_{FN} = x_{FN} H(ID_{FN}||C_a||C_b||T'_{FN})$.
4) Packs and sends $\{ID_{FN}, C_a, C_b, T'_{FN}, \sigma'_{FN}\}$ to CC.

3.6   Aggregated Data Decryption

1) CC checks the validity of the timestamp T'_{FN}. If it is valid, CC verifies the signature:

$$e(\sigma_{FN}{}', P) = e(H(ID_{FN}||C_a||C_b||T_{FN}{}'), pk_{FN}) \tag{7}$$

2) If the signature is verified, CC receives $\{ID_{FN}, C_a, C_b, T_{FN}{}', \sigma_{FN}{}'\}$. Then, CC broadcasts C_a to SM_i in a group and informs SMs to provide the partial decryption key.
3) SM_i uses to calculate the partial decryption key:

$$D_i = x_i(C_a) = x_i \sum_{i=1}^{n} C_{i,a} \tag{8}$$

Then SM_i signs on D_i to get $i \ ||D_i||T_i'')$.

4) SM_i sends $\{ID_i, D_i, T_i, \sigma_i{}''\}$ to FN, which will forward $\{ID_i, D_i, T_i{}'', \sigma_i{}''\}$ to CC.
5) CC checks the validity of the timestamp $T_i{}''$. If it is valid, CC performs the batch verification operation on signatures of n SMs:

$$e(\sum_{i=1}^{n} \sigma_i{}'', P) = \prod_{i=1}^{n} e(H(ID_i||D_i||T_i{}''), pk_i) \tag{9}$$

6) If the signature is verified, CC can obtain the following equation concerning the aggregated plaintext M:

$$M \cdot P = C_b - \sum_{i=1}^{n} D_i$$

$$= (a_1 \sum_{i=1}^{n} m_{i1} + a_2 \sum_{i=1}^{n} m_{i2} + \cdots + a_w \sum_{i=1}^{n} m_{iw} + b_1 S_{N_1} + b_2 S_{N_2} + \cdots + b_k S_{N_k}) \cdot P \tag{10}$$

Since M is a short message, M can be recovered by (10) according to Pollard's Lambda algorithm [20]. The total electricity consumption of the same type of all users M_j, and the number of users with m_i in the range $[R_u, R_{u+1})$ S_{N_u} can be recovered similarly. First, we consider the recovery of S_{N_u}:

Let $u = 1, 2,..., k-1$, and put them into the formula in turn: $S_{N_u} = (M - M \bmod b_u)/b_u S_{N_u}, M = M - (b_u S_{N_u})$. Let $u = k$, thank to $q >> b_u > b_0 + b_{u-1} \cdot n$, $b_0 > \sum_{j=1}^{w} a_j \cdot n \cdot T$, so we can obtain:

$$M \bmod b_k = a_1 \sum_{i=1}^{n} m_{i1} + a_2 \sum_{i=1}^{n} m_{i2} + \cdots + a_w \sum_{i=1}^{n} m_{iw} + b_1 S_{N_1} + b_2 S_{N_2} + \cdots + b_{k-1} S_{N_{k-1}}$$

$$SN_k = (M - M \bmod b_k)/b_k S_{N_k}$$

Next, M_j can be recovered: $M_j = (M' - M' \bmod a_j)/a_j$. Assign j to w, thanks to

$$0 \le m_{ij} < T, \sum_{j'=1}^{j-1} a_j' \cdot n \cdot T < a_j, \sum_{j=1}^{w} a_j \cdot n \cdot T << q.$$

$$M' \bmod a_w = a_1 \sum_{i=1}^{n} m_{i1} + a_2 \sum_{i=1}^{n} m_{i2} + \cdots + a_{w-1} \sum_{i=1}^{n} m_{i(w-1)}$$

$$M_w = (M' - M' \bmod a_w)/a_w = \sum_{i=1}^{n} m_{iw}$$

Here, M_w is the total electricity consumption data of all users in the w-th dimension. Similarly, all the remaining M_j can be recovered, $j = 1, 2, \ldots, w - 1$.

4   Analysis

In this section, we analyze the security and performance of the proposed scheme.

4.1 Security Analysis

The analysis on the security includes the privacy, the integrity and the authentication.

Privacy. The privacy means the electricity consumption data of each user is known only to himself. Our scheme can resist attacks from external adversaries and internal adversaries.

We assume that the internal attackers are CC, FN, $SM_i (i = 1, 2, \cdots, n-1)$. They want to collude to obtain the electricity consumption data of SM_n from $(C_{n,a}, C_{n,b})$. To do this, colluders must first calculate $PK \cdot r_n = C_{n,a} \sum_{i=1}^{n} x_i$. However, they cannot obtain x_n, implying it is computationally impossible for colluders to obtain the electricity consumption data of SM_n. Similarly, external adversaries cannot infer a user's electricity consumption data from $(C_{i,a}, C_{i,b})$ because they cannot obtain x_i.

Besides, entities can check the timestamp to resist malicious attackers and so our scheme can resist the replay attack.

Integrity. A malicious attacker may tamper with messages sent by the user. If the message is tampered with, it can be detected by a receiver. For example, when FN receives a message $\{ID_i, C_{i,a}, C_{i,b}, T_i', \sigma_i'\}$ from SM_i, FN checks ID_i and T_i', he verifies the integrity of the message by checking if $e(P, \sum_{i=1}^{n} \sigma_i') = \prod_{i=1}^{n} e(H(ID_i\|C_{i,a}\|C_{i,b}\|T_i'), pk_i)$ holds, any modification would not cause the equality hold. Similarly, the integrity of messages sent by FN can be verified by CC.

Authentication. Authentication guarantees the legitimacy of users. Our scheme ensures the legality of the users by the signature.

In the authentication phase, $(ID_i, T_i, \sigma_i, pk_i)$ can be verified by CC, and ID_i is included in $(ID_i, T_i, \sigma_i, pk_i)$. The signature $\sigma_i = x_i \cdot H(ID_i\|T_i)$ of SM_i is generated with the corresponding private key x_i. Since the adversary does not have the private key X_i, it cannot forge a verified $(ID_i, T_i, \sigma_i, pk_i)$. Therefore, our scheme can guarantee that all users are legal.

4.2 Performance and Comparison

We mainly analyze the performance of our scheme from the computation cost and the communication cost. Assume there are n SMs and each SM has w-dimensional electricity consumption data. Our experiments were conducted on a computer configured with 64-bit Windows 10 operating system, 2.40 GHz Intel Core, i5 CPU and 16 GB RAM.

Computation cost. We use the JPBC-based cryptographic library [22] to calculate the time cost. We compare the computation cost of our scheme with schemes [13, 18]. In our scheme, the computation cost of point addition operation on the elliptic curve group is ignored, and we only consider bilinear pairing operation T_{pair}, point multiplication operation T_{Mul} on elliptic curve, multiplication operation T_{mul} under Z_q, map-to-point hash operation T_H, exponential operation T_e. And scalar multiplication operation T_{ECC} under ECC and ElGamal encryption operation T_{enc} are considered. The computation cost of $T_{pair}, T_{mul}, T_{mul}, T_H, T_e, T_{ECC}$ and T_{enc} are 5.21 ms, 0.25 ms, 0.06 ms, 0.08 ms, 1.12 ms, 2.15 ms and 4.31 ms.

In our scheme, the total time cost of each SM and FN is $5T_{Mul} + (w + 2)T_{mul} + 2T_H = (0.06w + 1.53)$ ms and $(n + 1)T_{pair} + (n + 1)T_H + T_{mul} = (5.29n + 5.35)$ ms respectively. While in Zuo et al.'s scheme [13], the computation cost of each SM and the aggregator is $T_{enc} + wT_{mul} + 2T_e + T_H = (0.06w + 6.63)$ ms and $(n + 1)T_{pair} + (n + 1)T_H + (4n - 1)T_{mul} + T_e = (5.53n + 6.17)$ ms. Based on ECC, in Ming et al.'s scheme [18], the computation cost of each SM and the aggregator is $4T_{ECC} + (w + 3)T_{mul} + 2T_H = (0.06w + 8.94)$ ms and $(7n - 1)T_{mul} + (2n - 1)T_H + (2n + 2)T_{ECC} = (4.88n + 4.16)$ ms.

Figure 2 is the comparison of the computation cost of a SM when $w = 5, 10, 15, 20, 25, 30$. From Fig. 2, the computation cost of a SM of our scheme is lower than that of schemes [13, 18]. For multi-dimensional data aggregation schemes [13, 18], the computation cost of a SM increases with the data dimension linearly. The computation cost of a SM of our scheme based on ECC is lowest.

Figure 3 is the comparison of the computation cost of an aggregator when $n = 10, 20, 30, 40, 50, 60, 70, 80$. The computation cost of point addition operation on the elliptic curve group is ignored, the computation cost of an aggregator is not related to w. From Fig. 3, the scheme [13] has the highest computation cost among the schemes [13, 18] and ours, and the computation cost of our scheme is in the middle position, which is only slightly higher than that of the scheme [18]. Because the scheme [13] is not based on ECC, and its long ciphertext leads to its high computation cost, while both our scheme and the scheme [18] are based on ECC and the short ciphertext leads to the low computation cost.

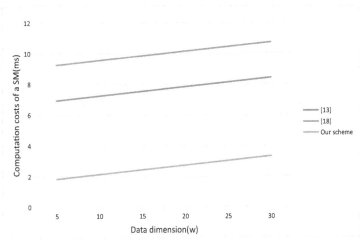

Fig. 2. Comparison of the computation cost of a SM.

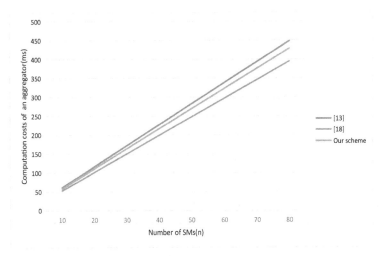

Fig. 3. Comparison of the computation cost of an aggregator.

Communication cost. The communication cost mainly consists of the communication cost from SMs to FN and FN to CC. Let G, G_T, Z_q^*, Z_N^* be 160 bits, 320 bits, 160 bits and 2048 bits, the length of the output value of hash function is 160 bits, and the length of identity and timestamp is 32 bits.

In our scheme, the communication cost from SM_i to FN and FN to CC is $(928n)$ bits and 928 bits. In the scheme [13], The communication cost from SM_i to the aggregator is $(32 + 160 + 160 + 32 + 320)n + (32 + 160 + 32 + 320)n = (704n + 544n) = (1248n)$ bits and the communication cost from the aggregator to the control center is also 1248 bits. In the scheme [18], the communication cost from SM_i to the aggregator is $(160 + 160 + 32 + 160 + 160 + 32)n = (704n)$ bits and the communication cost from the aggregator to the control center is 704 bits.

Figure 4 shows the relationship between the communication cost of SM-to-Aggregator and the number of SMs. The communication cost of our scheme is lower than that of the scheme [13], and slightly higher than that of the scheme [18]. This is because our scheme is based on ECC, which has the characteristics of the short key length and the low communication cost. Figure 5 shows the comparison of the communication cost of Aggregator-to-CC. The communication cost of our scheme is in the middle among schemes [13, 18].

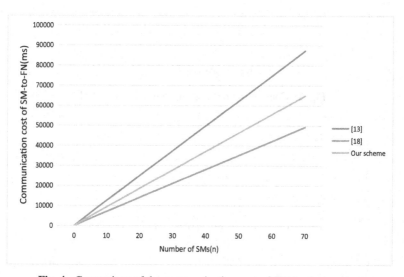

Fig. 4. Comparison of the communication cost of SM-to-Aggregator.

Comparison. In this section, schemes [4, 12, 13, 14, 18] are mainly compared with our scheme, from multi-dimensional data, integrity, dependence on TTP, resisting the replay attack and collusion-resistance. In the scheme [4], the control center can only analyze one-dimensional data. The scheme cannot guarantee the authentication and the integrity of the scheme. The scheme [14] relies on a TTP to generate blind factors. When internal participants collude, the user's electricity consumption data can be inferred. Similarly, the scheme [18] suffers from the same problem as the scheme [14], but the scheme [18]

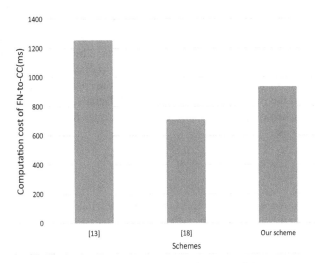

Fig. 5. Comparison of the communication cost of Aggregator-to-CC.

can perform multi-dimensional data analysis. The scheme [12] does not rely on a TTP, but when internal participants collude, colluders can infer the electricity consumption data of a user and the scheme does not consider the integrity. Although the other security features of the scheme [13] are satisfied, the scheme [13] cannot resist the replay attack in the authentication phase. The comparison results are shown in Table 1.

Table 1. Performance Comparison of Different Schemes

	[4]	[12]	[13]	[14]	[18]	Ours
Multi-dim	×	√	√	×	√	√
Integrity	×	√	√	√	√	√
Replay-resistance	×	×	×	√	√	√
No TTP	×	√	√	×	×	√
Collusion-resistance	×	×	√	×	×	√

5 Conclusion

In this paper, we propose a privacy-preserving multi-dimensional data aggregation scheme without a TTP in the smart grid. Before the data aggregation, SMs, FN and CC perform intra-group authentication. After the authentication, SM uses the group public key to encrypt the electricity consumption data. The super increasing sequences are used to keep the multi-dimensional data structure before and after the encryption, so that CC could obtain the specific number of users in a certain electricity consumption range and the total electricity consumption data of all users with the same type of the electricity consumption data. Therefore, CC can carry out fine-grained analysis on the electricity consumption data. All the keys are generated by the entity itself without the participation of the TTP, which avoids the security risk caused by the TTP being attacked. Besides, the collusion attack from some users, the control center and the aggregator is resisted, and the replay attack is also resisted. Detailed system analysis shows the security of our scheme, and the experiment demonstrates the computation cost and the communication cost.

References

1. Zhan, Y., Zhou, L., Wang, B., Duan, P., Zhang, B.: Efficient function queryable and privacy preserving data aggregation scheme in smart grid. IEEE Trans. Parallel Distrib. Syst. **33**(12), 3430–3441 (2022)
2. Wu, L., Xu, M., Fu, S., Luo, Y., Wei, Y.: FPDA: fault-tolerant and privacy-enhanced data aggregation scheme in fog-assisted smart grid. IEEE Internet Things J. **9**(7), 5254–5265 (2021)
3. Singh, P., Masud, M., Hossain, M.S., Kaur, A.: Blockchain and homomorphic encryption-based privacy-preserving data aggregation model in smart grid. Comput. Electr. Eng. **93**, 107209 (2021)
4. Li, S., Xue, K., Yang, Q., Hong, P.: PPMA: privacy-preserving multisubset data aggregation in smart grid. IEEE Trans. Industr. Inf. **14**(2), 462–471 (2017)
5. Zhang, W., Liu, S., Xia, Z.: A distributed privacy-preserving data aggregation scheme for smart grid with fine-grained access control. J. Inform. Secur. Appl. **66**, 103118 (2022)
6. Lu, R., Liang, X., Li, X., Lin, X., Shen, X.: EPPA: an efficient and privacy-preserving aggregation scheme for secure smart grid communications. IEEE Trans. Parallel Distrib. Syst. **23**(9), 1621–1631 (2012)
7. Lu, R., Alharbi, K., Lin, X., Huang, C.: A novel privacy-preserving set aggregation scheme for smart grid communications. In: 2015 IEEE Global Communications Conference (GLOBECOM), pp. 1–6 (2015)
8. Boudia, O.R.M., Senouci, S.M., Feham, M.: Elliptic curve-based secure multidimensional aggregation for smart grid communications. IEEE Sens. J. **17**(23), 7750–7757 (2017)
9. Shen, H., Zhang, M., Shen, J.: Efficient privacy-preserving cube-data aggregation scheme for smart grids. IEEE Trans. Inf. Forensics Secur. **12**(6), 1369–1381 (2017)
10. Anany, L.: Transform-and-conquer. Introduction to the Design and Analysis of Algorithms, 225–228(2002)
11. Chen, Y., Martínez-Ortega, J.F., Castillejo, P., López, L.: A homomorphic-based multiple data aggregation scheme for smart grid. IEEE Sens. J. **19**(10), 3921–3929 (2019)
12. Zeng, Z., Wang, X., Liu, Y., Chang, L.: MSDA: multi-subset data aggregation scheme without trusted third party. Front. Comp. Sci. **16**, 1–7 (2022)
13. Zuo, X., Li, L., Peng, H., Luo, S., Yang, Y.: Privacy-preserving multidimensional data aggregation scheme without trusted authority in smart grid. IEEE Syst. J. **15**(1), 395–406 (2020)
14. Fan, C.I., Huang, S.Y., Lai, Y.L.: Privacy-enhanced data aggregation scheme against internal attackers in smart grid. IEEE Trans. Industr. Inf. **10**(1), 666–675 (2013)
15. Bao, H., Lu, R.: Comment on privacy-enhanced data aggregation scheme against internal attackers in smart grid. IEEE Trans. Industr. Inf. **12**(1), 2–5 (2015)

16. He, D., Kumar, N., Zeadally, S., Vinel, A., Yang, L.T.: Efficient and privacy-preserving data aggregation scheme for smart grid against internal adversaries. IEEE Trans. Smart Grid **8**(5), 2411–2419 (2017)

17. Bao, H., Lu, R.: A new differentially private data aggregation with fault tolerance for smart grid communications. IEEE Internet Things J. **2**(3), 248–258 (2015)

18. Ming, Y., Zhang, X., Shen, X.: Efficient privacy-preserving multi-dimensional data aggregation scheme in smart grid. IEEE Access **7**, 32907–32921 (2019)

19. ElGamal, T.: A public key cryptosystem and a signature scheme based on discrete logarithms. IEEE Trans. Inf. Theory **31**(4), 469–472 (1985)

20. Boneh, D., Goh, E.-J., Nissim, K.: Evaluating 2-DNF formulas on ciphertexts. In: Kilian, J. (ed.) TCC 2005. LNCS, vol. 3378, pp. 325–341. Springer, Heidelberg (2005). https://doi.org/10.1007/978-3-540-30576-7_18

21. Boneh, D., Lynn, B., Shacham, H.: Short signatures from the weil pairing. In: Boyd, C. (ed.) ASIACRYPT 2001. LNCS, vol. 2248, pp. 514–532. Springer, Heidelberg (2001). https://doi.org/10.1007/3-540-45682-1_30

22. De Caro, A., Iovino, V.: jPBC: Java pairing based cryptography. In: 2011 IEEE Symposium on Computers and Communications (ISCC), pp. 850–855. IEEE (2011)

Using Micro Videos to Optimize Premiere Software Course Teaching

Lixiang Zhao, Xiaomei Yu$^{(\boxtimes)}$, Wenxiang Fu, Qiang Yin, Haowei Peng, and XiaoTong Jiao

Shandong Normal University, Jinan, China
yxm0708@126.com

Abstract. The rapid development of modern educational technology gives rise to novel forms in teaching reform. The innovative teaching methods are booming to settle the issues in secondary vocational courses, such as the lack of motivation and initiative in learning. In this paper, we present the design and application of micro videos in Premiere Software course based on eye tracking technology. Specifically, we investigate the feasibility of incorporating micro videos into course learning firstly, and then explore effective micro videos design strategies in Premiere Software learning based on eye tracking technology. Finally, we apply the micro videos in class teaching of Premiere Software for secondary vocational students, and the teaching evaluation is conducted after a semester of course learning. The evaluate results demonstrate that the micro videos based course teaching is effective to arouse learning interest of secondary vocational students, and further improve efficiency in their Premiere Software learning.

Keywords: Micro videos · Eye tracking · Teaching effectiveness · Vocational education

1 Introduction

The rapid development of information technology has had a profound impact on education, and has given rise to new teaching patterns. Information technology teaching is a mode of changing teaching mode and improving teaching efficiency. Many schools are still following the traditional teaching mode, with problems such as low interest, outdated teaching concepts and outdated teaching mode. The application of micro videos in education is beneficial to enriching teaching patterns and improving teaching efficiency. The micro videos brings a new mode to teachers and students, it reflects students' subject status and stimulates students' enthusiasm.

With the advantage of intuitive, vivid and easy to spread, the micro videos have attacked the attention of many teachers. For a teacher, it is difficult to provide each student with proper learning sources according to his personalized learning needs in traditional classroom learning scenarios. Therefore, it is necessary for a teacher to give well-designed micro videos and provide timely assistance for the students to pick up the key points in their learning.

© The Author(s), under exclusive license to Springer Nature Singapore Pte Ltd. 2024
J. Vaidya et al. (Eds.): AIS&P 2023, LNCS 14510, pp. 92–105, 2024.
https://doi.org/10.1007/978-981-99-9788-6_8

Visual attention is a theory that explains how the human visual system selects, processes, and allocates attention resources. It is crucial in understanding human perception, cognition, and decision-making processes [1]. Eye tracking technology records individual eye movements and gaze positions during various tasks [2]. Researchers use this technology to study the cognitive processes behind individual visual experiences [3]. The application of eye tracking technology in multimedia learning research is based on the eye mind hypothesis, which suggests that visual attention is a proxy for mental processing and reflects individual cognitive strategies. When learners interact with different forms of expression (such as text, graphics, animation, video) in a multimedia learning environment, eye tracking technology assesses attention allocation and cognitive strategies of multimedia learners [4].

The purpose of this study is to explore a teaching mode that integrates micro videos into the Premiere Software course, aiming to enhance its teaching effectiveness student's learning efficiency. By adding additional content, micro videos enable students to acquire more comprehensive knowledge in the course. The contributions and works of this paper are summarized as follows:

- In this paper, we investigate the feasibility of incorporating micro videos into course learning firstly.
- Then explore effective micro videos design strategies in Premiere Software learning based on eye tracking technology.
- Finally, we apply the micro videos in class teaching of Premiere Software for secondary vocational students, and the teaching evaluation is conducted after a semester of course learning.

2   Related Work

2.1   Micro Videos

Micro video is a 5–10 min teaching video clip recorded by teachers under the guidance of certain teaching ideas and learning theories. According to teaching rules, teachers divide teaching objectives and course content into knowledge units. The concepts of mini course and mini lecture are also related to micro videos. The term "mini course" was proposed by American scholar Borg. It mainly focuses on establishing miniature courses and refers to small modular courses with themes [5,6]. American scholar David Penrose proposed the concept of "mini lecture". It generally appears in the form of speeches, focusing on the use of scenarios and the design of activities, tools, and templates.

David Penrose also proposed a five-step process for creating micro videos: listing core concepts; providing contextual knowledge; recording and producing 1–3 min teaching video programs; designing after-school tasks for autonomous learning and exploratory learning; and uploading teaching videos and course tasks to the course management system [7]. Many researchers have conducted research on micro videos. Wikipedia defines micro videos as using mobile devices and online learning methods to form actual teaching content [8]. Shieh believes that micro videos are a special speech that uses video as a carrier, which helps learners improve their classroom participation through short videos and students' existing knowledge level [9].

The application of micro videos in classroom teaching can take two forms: flipped classroom and mixed teaching. The flipped classroom, initially proposed by Aaron Sams, is a teaching method that believes it can effectively assist students with slower learning progress. By incorporating micro videos, students can access course content that they may have difficulty understanding at any time, thereby enhancing their learning outcomes. Zichao Chen has applied micro videos to flipped classroom teaching and, based on his years of teaching experience, has summarized the application strategies for using micro videos. These strategies aim to help students preview, learn in class, review, and evaluate after class [10].

The inclusion of micro videos allows teaching to truly cater to every students. By viewing micro videos multiple times, students can overcome time and space constraints, keeping up with the teaching progress. Micro videos are also being implemented in various specific disciplines. Xiaojiao Chen suggests focusing on the issues present in classroom teaching and incorporating micro videos into specific teaching processes. She further analyzed the effectiveness of micro videos in continuously improving teaching [11]. Huazhong Ding's work involves using micro videos in the teaching of information technology subjects in middle schools. By analyzing the teaching process and its effectiveness, he has proposed better applications of micro videos in information technology classroom teaching [12]. The application of micro-videos makes teaching more personalized and helps to improve the learning outcomes of students.

Micro videos design strategies have become a hot topic of interest for many researchers. After analyzing the award-winning works from the first China Micro Videos Competition, Yonghua Wang proposed suggestions from thee aspects: teaching design, language expression, courseware and video production [13]. Xianzhong Cao used experimental methods to explore the impact of different software operation micro videos on learning outcomes, and proposed design suggestions from the perspective of optimizing learning outcomes [14]. Tiesheng Hu proposed the ADDIE model for designing micro videos, which has been widely recognized and referenced. This model not only focuses on video effects, but also emphasizes students' learning experiences [15]. Xiangzeng Meng proposed using a constructivist teaching mode to design micro videos. This mode leading learners to construct knowledge from teaching cases to teaching applications, and comprehensively improve teaching effectiveness [16]. Jianxia Cao focused on studying the impact of the proportion of teachers appearing in micro videos on students' learning outcomes, and proposed micro videos design strategies to achieve better teaching outcomes [17]. These studies provide strong theoretical support and practical guidance for the application of micro-videos in classroom teaching.

2.2   Eye Tracking Technology

Eye movements serve as behavioral indicators that relate to visual and cognitive processing, assisting in understanding attention allocation [18]. When readers engage with complex information, they are required to read and process text, as well as browse illustrations, to derive meaning. If the two sources of information are distant from each other, this process may impose cognitive demands [19]. Eye tracking technology present an opportunity to test where people look when integrating text and images [20]. Eye tracking technology primarily consists of two measurement methods: Gazing and Scanning. Gazing refers to the eye's scanning state at a specific point, while scanning denotes the rapid movement of the eye between fixations. This indicates a change in the visual focus of attention [21].

Eye tracking technology captures fine-grained data of students when they watch micro videos, thereby promoting the improvement of multimedia resource design strategies [22]. Scholars believe that using eye movement technology in micro video learning captures fine-grained data, thereby establishing a connection between micro video design strategies and students' learning processes through eye movement data [23]. Sasha employed three eye movement parameters: the total number of fixation points, total fixation time, and average fixation time to explore the attractiveness of educational video content to learners [24]. Xiaoming Cao analyzed the relationship between student's learning effectiveness and the difficulty of learning materials by using the number of fixation points. They explored the difficulty of students extracting information through fixation time, and investigated students' concentration when watching educational videos using visual loss time [25]. Numerous experimental results demonstrate that eye tracking technology can map learners' cognitive processes during learning. Eye tracking technology allows for the measurement of learners' learning effectiveness at the level of learning cognition [26].

3   Method

3.1   Micro Videos Feasibility Questionnaire

We aim to explore the current status of the Premiere Software Course and the potentiality of incorporating micro videos into it through a questionnaire. The research participants are third-year students majoring in digital media at a vocational school. These students have already studied the Premiere Software course in their second year, thus, they have a comprehensive understanding of the curriculum model and the teaching process of the original course. In order to understand the course situation as comprehensively as possible, we distributed the survey questionnaire to them. The questionnaire used a paper format, which is more suitable for the research participants who are students in school, compared to an online questionnaire. A total of 105 third-year students received the questionnaires. The students filled out the questionnaire with a pen, and returned it after completion.

The questionnaire is divided into three parts. The first part primarily focuses on surveying students' basic situation, with a particular emphasis on their level of enthusiasm and learning motivation for Premiere Software Course. The second part aims to understand the learning situation of the course, examining students' needs and their opinions regarding the teaching content, teaching methods, and evaluation strategies. The third part investigates students' familiarity with micro videos and their views on integrating micro videos into the course. The questions provide 3–5 options, allowing students to choose the most suitable option. The data collected is processed viapaper and pencil, ultimately resulting in the calculation of the questionnaire's outcomes.

3.2 Eye Tracking Experiment

In this section, we aim to investigate the effects of micro videos with and without subtitles on students' learning through eye movement experiments. The eye tracking experiment involved 10 students, all of whom were sophomore students studying digital media technology applications at a vocational school. During the experiment, the students were randomly divided into two equal groups. Prior to participating in the experiment, the students acquired some professional knowledge and were capable of understanding the micro video content well. This reduced the impact of other factors on the students' cognitive load. The participants' naked eye or corrected vision was normal, with no astigmatism, color blindness, or color weakness. There was no significant difference in vision between their left and right eyes. They had not participated in similar eye movement experiments and were not familiar with the implementation principles of eye tracking technology.

The experimental equipment utilizes the Eyeso Ex150 remote sensing eye tracking system, with a sampling rate of 150FPS. The experimental data recording and analysis utilize its built-in Eyeso Studio 3.3 eye movement experimental design and data analysis system. The stimulus materials for this experiment are two pre-recorded micro videos groups. The content of this micro videos originates from Premiere Software Course. To better control the experimental variables, the content of the two micro videos groups is identical, except for the presence or absence of subtitles.

This experiment recorded and analyzed the distribution of students' visual attention when watching micro videos. It also explored the impact of subtitles on students' cognitive load. The experimental steps are as follows:

- Introduce the experimental theme. Researchers introduced the experimental topic to the participants and studied the attention distribution of students when watching micro videos. The experimental method involved dividing students into two groups, then having them watch micro videos with subtitles or without subtitles.
- Preparation for the experiment. Researchers entered the personal information of the subjects into the system before the experiment began, such as age, gender, and other basic personal information.

- Experimental calibration. Researchers required the subject to sit at a distance of about 60cm from the screen, so that the eye tracker could capture the student's pupils well. The calibration process began when the subject adjusted their sitting posture to the best position. The experiment could only begin When the subject's eye tracker calibration error was less than or equal to 40, indicating that the calibration requirement was meet.
- Experimental process. Students were briefed on the content of the micro video and provided with a computer to watch it. The duration of the micro video was 10 min. Students were required to carefully study the content within the micro video.
- Save the data. After the experiment completed and the micro video played, researchers save the subject's eye movement data.

3.3 Online and Offline Blended Teaching

The blended teaching method refers to a teaching strategy that combines online and offline methods to accomplish teaching tasks within a specific time frame. Vocational education focuses not only on students' knowledge acquisition but also their vocational skills development. By adopting a combined approach of online and offline teaching, teachers can introduce job-required through online courses and teach relevant theoretical knowledge through offline courses.

3.4 Evaluation of Teaching Effectiveness

We access the effectiveness of the course by students' final grades and course satisfaction questionnaires. Teachers utilize an improved teaching method to teach students after evaluating its effectiveness. The questionnaire is used to investigate changes in students' abilities and their level of satisfaction with the course. The questionnaire employs the Likert scale, where responses are recorded on a scale of 1 to 5, corresponding to five dimensions: very disagree, disagree, general, agree, and very agree. These represent the degree of recognition respondents have towards the course. Higher levels indicate higher student satisfaction with the course and improvements in their abilities. The teaching effectiveness questionnaire was distributed to second-year digital media students who have studied the reformed Premiere Software course in this semester. The format of the questionnaire is in paper form.

The questionnaire is divided into two sections. The first section analyzes changes in students' ability. This part investigates their improvements in digital audio and video editing skills, learning interest, self-learning capacity, independent thinking, communication and cooperation skills, innovation abilities, and problem-solving skills. The second section of the questionnaire evaluates students' satisfaction with the reformed course. The data was collected through paper and pen methods. We utilize IBM SPSS Statistics 25 to analyze the reliability and validity of the questionnaire. We compile the questionnaire responses and count the number of 1–5 levels for each question. Afterward, we calculate the mean and standard deviation, followed by a specific analysis.

4   Results

4.1   The Feasibility of Micro Videos

The questionnaire was distributed to 105 students, and 102 valid questionnaires were received, with the effective rate of 97.1%. The first part of the questionnaire mainly includes two questions. Regarding the students' interest in the course, 61% of the them expressed great interest. About 21% of the students are interested, and 18% of the students are generally interested. As shown in Fig. 1.

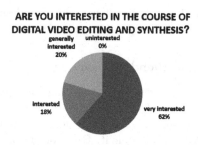

Fig. 1. Statistical chart of course preference

In terms of course learning motivation, the data shows that 49% of the students attribute their motivation to employment accounts. 21% of the students choose entrance exams. 10% of the students select hobbies and others interests as their motivation. These statistical results are presented in Fig. 2.

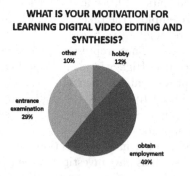

Fig. 2. Statistical chart of course preference

The second part of the questionnaire comprises four questions. The third question pertains to whether the current course meets the students' needs. Of the students, 15.7% are satisfied, 64.7% are basically satisfied. In addition, and 19.6% are not satisfied. Regarding the improvement of the course's teaching content, 15.7% of the students wish to maintain the current situation. 24.5% of students hope for a deeper understanding of the content, 27.5% desire a more practical approach, and 32.3% wish for a more operational course. As for the teaching method, 41.1% of students opt for the cooperative inquiry mode. 23.5% choose self-learning tutoring, 22.5% prefer lectures and 12.7% want to maintain the status quo. In terms of course evaluation methods, 7.8% of the students choose to stick to the current system. 62.8% of the students hope for more diversity and 29.4% advocate for a focus on process evaluation.

In the third part of the questionnaire, students were asked about their exposure to micro videos. A total of 36.3% of students reported frequently coming into contact with micro videos, 45.7% occasionally came across them, and 17.1% had never been exposed to them. When asked about their interest in incorporating micro videos into the course, and overwhelming 80.4% of students expressed interest. Another 10.8% were generally interested, while only 8.8% of students were not interested.

According to the questionnaire survey on the feasibility of micro videos, it can be analyzed that the majority of students are interested in the Premiere Software course. Therefore, it is meaningful to reform the Premiere Software course. It is possible to incorporate micro videos into the course, as 80% of students have been exposed to them. Most students are interested in Premiere Software course that incorporates micro videos. Among all the motivational options for learning courses, employment was selected by the majority of students. Thus, it is important to include knowledge and skills related to the corresponding position in the course. Only 1/6 of the students chose to maintain the current teaching content, while the majority students hope for an improvement, practicality, or depth in the content. In terms of teaching mode, more than half of the students prefer the cooperative exploration mode or self-study tutoring teaching mode. Most students prefer diverse and diverse evaluation methods.

4.2 Micro Videos Design Strategies

When students watch micro videos, they tend to automatically focus on specific areas of the screen while unintentionally neglecting others. Eye tracking data offers a three-dimensional representation of students' behavior while watching videos, which is of great significance for the design of micro videos. The key question is how to create micro videos that enable students to grasp the main concepts and reduce their cognitive load. We analyze students' attention distribution while watching micro videos through eye movement heat maps to gain insights into the issue.

A heat map is a visual representation of a subject's eye movement data. It directly displaying the area of focus on the interface. It use color to indicate the student's gaze, with strong interactions occurring in the red, orange and yellow areas. The moderate interactions occurring in the green areas, and slight interactions in the blue areas. This experiment analyzes students' heat maps to study their attention distribution while watching micro videos. The results show that students' attention is more dispersed when watching micro video without subtitles. When students watch micro videos with subtitles, their attention is primarily focused on the subtitle area of the video. The heat map of the micro video without subtitles is presented in Fig. 3, and the heat map of the micro video with subtitles is shown in Fig. 4.

Eye tracking technology provides an objective and unique method for studying users' attention to micro videos interfaces, contributing to the design's effectiveness. According to the eye movement experiment's heat map, students' attention is more distracted when micro lesson videos are designed without subtitles. This requires students to exert extra cognitive load to grasp the learning focus of the micro videos. When micro videos include subtitles, students' attention is primarily focused on the subtitle area, making it easier for them to understand the content and reducing cognitive load. We should therefore design micro videos with subtitles to synchronize auditory and visual materials, facilitating better understanding for students. Compared to micro videos without subtitles, those with subtitles are more conducive to students' learning.

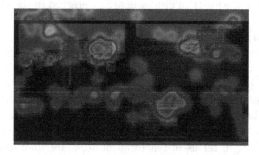

Fig. 3. hotspot map of untitled micro videos design

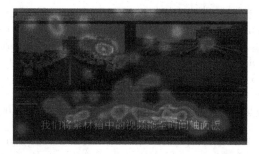

Fig. 4. hotspot map of subtitled micro videos design

Multimedia learning theory also provide insights into micro videos design strategies. According to the method of reducing cognitive load in multimedia learning theory proposed by Richard E. Mayer, the design strategies of micro videos can be summarized into three aspects: controlling redundancy, marking to reduce burden, and proximity effect reduction. Controlling redundancy refers to the current trend in micro videos design and production towards popularization and entertainment. Designers often incorporate elements such as emoticons and special sound effects that are not directly related to the learning content. These measures may potentially distract students from the key learning content.

Therefore, when creating and manufacturing micro videos, we should eliminate irrelevant content and emphasize the key information to ensure that the micro videos are concise and clear. This guides learners to perform the cognitive processing necessary for learning, and minimize the processing and input of additional information while studying the micro videos. The marker to reduce burden refers to highlighting key phrases with colors, boldness, and circles, which helps learners locate the key content quickly and establish cognitive causal relationships. Although create concise and clear demonstration materials can be challenging in a real-life learning environment, arrows can still be used to mark important information in images. The proximity effect reduction refers to the simultaneous presentation of visual and auditory materials in micro videos design, which aims to minimize learners' cognitive load.

4.3   Micro Videos in Practice

After confirming the feasibility of micro videos, the next step is to incorporate them into the actual teaching process. We have adopted a blended learning approach, combining online and offline elements. In class, the teacher explains the basic knowledge, key concepts, high-quality case analysis, and overall project arrangement and guidance for the Premiere software course. Micro videos are used to explain the workflow and various related standards of the film and television industry, connecting with enterprise positions to enhance students' professional abilities and industry literacy. In order to better foster students' practical skills, a project practice stage is arranged to provide hands-on opportunities while teaching theoretical knowledge.

The project practice stage is divided into two stages: creative filming and film production. After shooting the material, it is edited into a film. During the practical stage, micro videos are utilized to provide students with guidance on script creation, paperwork and directing, drawing and framing, photography, and other knowledge. These skills are very highly beneficial for students to write and develop scripts, as well as shoot videos. Due to the repeatable nature of micro videos, students can watch them multiple times according to their needs, which enables them to meet their individualized learning requirements.

When innovating the teaching method, we have also update the content of the Premiere Software course. Regarding the teaching content, we will merge talent development plans, course textbooks, ideological and political aspects, industry job requirements, and elements from students' daily lives into eight open teaching projects. These eight projects include the Charming Cities, Wonderful Campus Life, Attractive China, I and Communist Youth League, Exercising

Health, Craftsmanship Spirit, Campus Civilization Starting From Me, and I and the School. The Charming City project theme revolves around the city where the students are located. While the Wonderful Campus Life project focuses on the campus life that the students are most familiar with.

According to the teaching plan, the implementation of the Amazing China project aligns with the National Day, making its content more festive. As students join the Communist Youth League at their current grade, we have chosen the project themes of I and the Communist Youth League. The Exercising Health project's theme is related to the campus sports event, and the Craftsman Spirit project aligns with the vocational school's educational characteristics. The Campus Civilization Starting From Me project aims to cultivate students responsibility for maintaining campus civilization. By integrating student' living cities, campus life, sports and health related matters into teaching projects, we aim to stimulate students' interest in learning. In class, teachers arrange textbook knowledge points and appropriately explain special effects, making the course more practical and operation-oriented based on the project themes. Students personally participate in this process, enabling them to have a deeper memory and better grasp of course knowledge.

4.4   Course Effectiveness Assessment

Following a semester of practice, we access the effectiveness of course reform based on students' final grades and a course effectiveness survey questionnaire. The final score of a student consists of two components: the scores of eight items and the final examination scores. The project scores are determined by student self-evaluation, peer evaluation, and teacher assessment. The final scores of the students are summarized in Table 1. There are 42 students in the class, with a highest score being 98 and the lowest score being 65. The average score is 81.48, and the standard deviation is 7.29. Overall, the distribution of scores is relatively average, and the difference between scores falls within a reasonable range. The majority of students have achieved favorable outcomes.

Table 1. Student Performance Statistics Table

Score range	Proportion	standard deviation
90–100	11.9%	7.29
80–89	47.6%	
70–79	33.3%	
60–69	7.1%	
0–59	0	

The course effectiveness survey questionnaire employs the Likert scale method, where the options for each question are scored from 1 to 5. The

1–5 points correspond to five grades: very disagree, disagree, neutral, agree, and very agree. We utilize SPSS software to analyze the reliability and validity of the questionnaire. The reliability of the questionnaire is 0.905, which is greater than 0.8. The reliability indicates that the questionnaire design has good stability and the survey results are stable. The KMO (Kaiser-Meyer-Olkin) value of the questionnaire is 0.859, which is greater than 0.7. The significance (sig) value is 0, which is less than 0.05. This indicates that the questionnaire results are accurate and effective, providing reliable data for further analysis.

The course effectiveness survey questionnaire also reveals that the standard deviation of each question is approximately 1, indicating a relatively small fluctuation in the responses. The average score for each question is around 4, which suggests that the majority of students agree that combining micro videos with the course has led to an improvement in their abilities. Furthermore, students are satisfied with the teaching effectiveness of the course after the reforms. They believe that the existing course teaching methods and content arrangements are more suitable and effective. Overall, these findings indicate that the integration of micro videos and the course has been well-received and has positively impacted students' learning experiences.

5 Discussion

The integration of micro videos in the Premiere Software course in vocational schools is currently being implemented in only one class due to times and resource constrains. Although the current impact is evident, the research' scalability requires further verification through extensive practice before it can be widely promoted. Additionally, the teacher involved lacks sufficient teaching experience, and their instructional abilities need to be enhanced. Lastly, the micro video resources are not yet comprehensive, primarily due to hardware limitations.

There is considerable room for advancement in the realm of micro videos production. Moving forward, we will persist in investigating and enhancing the teaching design that incorporates micro videos. Moreover, we will still seek ways to boost the innovation and practicality of our courses through the implementation of micro videos. We aspire to attract a larger contingent of micro videos instruction. This team will continually explore and update our teaching resource library. We will also actively encourage the integration of micro videos into professional course instruction.

6 Conclusion

This article explores the potential of integrating micro videos into a Premiere Software course through a questionnaire survey. Building on this foundation, we have incorporated micro videos into the course and studied the design strategies of micro videos. Eye tracking technology was employed to analyze the effects of different design strategies on students' attention. Guided by multimedia learning theory, we have summarized design strategies for micro videos, including controlling redundancy, reducing label burden, and mitigating proximity effects. We have also investigate the optimal time for incorporating micro videos into course teaching. Furthermore, the integration of micro videos has led to the optimization of teaching content, mode, and evaluation.

In terms of the teaching mode, we have divided the process into three stages: classroom instruction, project practice, and error correction. Simultaneously, we have also reformed the teaching evaluation method. After a semester of implements these changes, we assessed the effectiveness of the course reform through students' final grades and a classroom teaching effectiveness questionnaire. The findings reveal that 93% of students earned satisfactory grades. Furthermore, students concur that their communication and collaboration with teachers and peers improved, as did their self-learning, independent thinking, and collaboration skills. Overall, the course instruction achieved the desired outcomes.

Acknowledgements. This work is supported by Shandong Provincial Project of Graduate Education Quality Improvement (No. SDYJG21104, No. SDYJG19171), the Key R&D Program of Shandong Province, China (NO.2021SFGC0104, NO.2021 CXGC010506), the Natural Science Foundation of Shandong Province, China (No. ZR2020LZH008, ZR2021MF118, ZR2022LZH003), the National Natural Science Foundation of China under Grant (NO. 62101311, No. 62072290), the Postgraduate Quality Education and Teaching Resources Project of Shandong Province (SDYKC2022053, S DYAL2022060), the Shandong Normal University Research Project of Education and Teaching (No.2019XM48), and Industry-University Cooperation and Education Project of Ministry of Education (No. 220602695231855).

References

1. Yang, Q., Jiang, Y., Shen, L.: A study on the influence of metaphorical rhetoric on the attention effect of online banner advertising. J. Manage. Eng. **37**(2), 48–59 (2023)
2. Carter, B.T., Luke, G.L.: Best practices in eye tracking research. Int. J. Psychophysiol. **155**, 49–60 (2020)
3. Zheng, Y., Song, X., Gao, G., Li, H.: New progress in the application of eye movement tracking technology in multimedia learning abroad: evidence from related research from 2016 to 2021. Expl. Educ. Sci. **41**(1), 96–102 (2023)
4. Gibaldi, A., Vanegas, M., Bex, P.J.: Evaluation of the Tobii ExeX Eye Tracking controller and Matlab toolkit for research. Behav. Res. Meth. **49**, 923–946 (2017)
5. Borg, W.: A microteaching approach to teacher education. Macmillan Educ. Serv. **1**, 256 (1970)
6. Lees, C.: Model of emotional intelligence mini lecture. BMC Neurol. **10**(1), 103 (2011)
7. Orgeman, D.J.: Mini-lessons in revision to support the writing growth of primary grade students. Journal of Management Engineering. Hamline University: School of Education and Leadership Student Capstone Theses and Dissertations (2008)
8. Wikipedia. http://en.wikipedia.org/wiki/Microlecture. Microlecture
9. Shieh, D.: These lectures are gone in 60 seconds. Chronicle Higher Educ. **55**(26), 1–13 (2009)

10. Chen, Z.: Research on teaching design of flipped classroom integrating micro class and MOOC. China Educ. Technol. **09**, 130–134 (2017)
11. Chen, X., Yang, Z., Yan, Y.: Design of chemical experiment microcourse based on primary teaching principles. Chem. Educ. **39**(6), 35–38 (2018)
12. Ding, H.: A Research on the Design, Production, and Application of Micro Video Teaching Resources in Teaching. Central China Normal University (2016)
13. Wang, Y., Ma, X.: Analysis and research on micro videos videos in primary and secondary schools in China - based on the analysis of winning videos in the first china micro videos competition. Teach Manage. **09**, 103–105 (2015)
14. Cao, X., Tong, S., Xu, S., Ma, H.: Experimental study on the effect of subtitle configuration on learning effectiveness in software operations Microcourses. China Educ. Technol. Equip. **18**, 41–44 (2018)
15. Hu, T.: What changes have micro courses brought to education. Inf. Technol. Educ. Prim. Sec. Sch. **02**, 84–86 (2018)
16. Meng, X., Liu, R., Wang, G.: Theory and practice of micro videos design and production. J. Dist. Educ. **32**(06), 24–32 (2014)
17. Cao, J., Fu, A.: The impact of teacher presentation in teaching videos on learners' social presence and learning outcomes. Mod. Educ. Technol. **27**(7), 75–81 (2017)
18. Henderson, J.M., Ferreira, F.: The interface of language, vision, and action: eye movements and the visual world. Scene Percept. Psycholinguists **1**, 1–58 (2004)
19. Holsanova, J., Holmberg, N., Holmqvist, K.: Reading information graphics: the role of spatial contiguity and dual attentional guidance. Appl. Cogn. PsyChol. **23**, 1215–1226 (2009)
20. Tabbers, H.K., Paas, F., Lankford, C., Martens, R. L.: Studying eye movements in multimedia learning. Understanding Multimedia Documents, pp. 169–184 (2008)
21. Van Gog, T.: Eye tracking as a tool to study and enhance cognitive and metacognitive processes in computer-based learning environments. In: International handbook of metacognition and Learning Teachnologies, pp. 143–156 (2013)
22. Scheiter, K., Gog, T.V.: Using eye tracking in applied research to study and stimulate the processing of information from multi-representational sources. Appl. Cogn. Psychol. **23**(9), 1209–1214 (2010)
23. Alemdag, E., Cagiltay, K.: A systematic review of eye tracking research on multimedia learning. Comput. Educ. **125**, 413–428 (2023)
24. Chen, X., Zheng, X., Sun, K., Liu, W., Zhang, Y.: Self-supervised vision transformer-based few-shot learning for facial expression recognition. Inf. Sci. **634**, 206–226 (2023)
25. Wang, T., Zheng, X., Zhang, L., Cui, Z., Xu, C.: A graph-based interpretability method for deep neural networks. Neurocomputing **555**, 126651 (2021)
26. Yin, Y., Zheng, X., Hu, B., Zhang, Y., Cui, X.: EEG emotion recognition using fusion model of graph convolutional neural networks and LSTM. Appl. Soft Comput. **100**, 106954 (2021)

The Design and Implementation of Python Knowledge Graph for Programming Teaching

Xiaotong Jiao, Xiaomei Yu[✉], Haowei Peng, Zhaokun Gong,
and Lixiang Zhao

School of Information Science and Engineering, Shandong Normal University,
Jinan, China
yxm0708@126.com

Abstract. With the continuous development of information technology in education, Python course has a pivotal position in the current information technology curriculum system. But there are the following problems in the process of teaching: it is difficult for students to construct meaningful knowledge; the teaching materials cannot effectively stimulate the interest in learning; content taught in the classroom cannot meet the education needs either. Therefore, it is urgent to construct a Python Knowledge Gragh (KG) that covers heterogeneous data from multiple sources and enables knowledge visualization. We constructed a Python KG based on the Neo4j graph database and interactive graph framework from the perspectives of innovating how to present teaching content, improve learning interest, and expand teaching resources. We demonstrated by designing teaching experiments that the Python KG could address the issues of information overload and passive learning by making Python more accessible, engaging, and personalized.

Keywords: Knowledge Graph · Python course · Teaching practice · Instructional design · Informatization

1 Introduction

With the rapid development of computer technology, network technology, multimedia technology and other information technology, globalization and informatization have become the new trend in the development of education nowadays. Informatization has become an inevitable requirement for educational reform and development, and it is also a necessary path to innovate and improve teaching methods, an important measure to improve teaching quality.

In 2016, the Ministry of Education issued the "13th Five-Year Plan for Education Informatization", proposing to "actively promote the integration of information technology and education" to accelerate the process of education informatization. In 2018, the Ministry of Education formulated and issued the "Education Informatization 2.0 Action Plan", comprehensively promoting the "Internet" +

© The Author(s), under exclusive license to Springer Nature Singapore Pte Ltd. 2024
J. Vaidya et al. (Eds.): AIS&P 2023, LNCS 14510, pp. 106–121, 2024.
https://doi.org/10.1007/978-981-99-9788-6_9

education process, adhering to innovation, and proposing to integrate emerging technologies into teaching [1]. At the beginning of 2023, Li Yongzhi, president of the Chinese Academy of Education Sciences, proposed in the Bluebook 2022 of China's Smart Education that in terms of education content, a digital Knowledge Graph (KG) should be established based on the systematic logical relationship of knowledge points, and the presentation of education content should be innovated [2]. Developing a KG for courses is a valuable initiative with significant implications for education and technology-enhanced learning [3]. A KG can help students navigate the vast amount of information available for lessons. It provides a structured and interconnected way to access learning materials, making the learning experience more intuitive and efficient. A KG, on the other hand, can be customized to suit individual learning needs. Learners can access content that matches their skill level, learning style, and goals, promoting a more personalized and effective learning journey. Therefore, the construction of KG according to the characteristics of the curriculum is a direction in which we can make drilling.

Python is an important course for middle- and high-level computer majors. The teaching goal of the Python course is to enable students to master the basic concepts, introductory statements, basic data types, functions, etc., of the Python programming language; to improve the student's ability to write programs independently; to cultivate students' computer thinking and to lay the foundation for students' future education or employment.

However, in terms of the actual teaching situation of Python courses, there are a series of problems: Firstly, Python teaching puts higher requirements on students' comprehension and imagination, making it difficult to teach and learn and is not conducive to students' meaningful construction of knowledge. Secondly, a large portion of the currently used textbooks and underclass tutorials present knowledge in the form of descriptive text, which is a relatively high information density and low acceptance for students, and it is easy for students to resist the learning of Python courses instinctively. Therefore, it is necessary to construct a KG with data from multiple sources. Adopting KG visualization technology to display abstract concepts in graphs can increase students' interest in learning, make students acquire knowledge more intuitively, generate strong motivation to learn, and change from passive learning to active learning.

In this study, we compile a KG that integrates data from various reliable sources to offer a comprehensive and up-to-date repository of Python-related information, resources, and concepts. Utilize KG visualization technology to create an interactive, user-friendly platform visually representing abstract Python concepts in graph form. The Python KG enables students to grasp complex Python topics more intuitively, making learning Python more accessible and less intimidating. Moreover, it can motivate students to become active learners by engaging with the KG, allowing them to explore Python concepts independently and at their own pace. We aim to provide students with personalized learning paths and recommendations within the KG to cater to individual skill levels and learning preferences. To verify the effectiveness of this design, we also applied it to teaching; the main work and innovations are as follows:

- Design: We build a Python KG based on the Neo4j graph database and Interactive Graph framework.
- Construction: We propose an information tool based on KG for teaching Python courses and design usage methods to assist teaching.
- Implementation and evaluation: We based on Python KG for instructional design and its application to teaching practice; through the design of teaching experiments to verify the effectiveness of Python KG, the questionnaire survey method was used to evaluate the teaching method of integrating KG.

2 Related Work

Knowledge graph is not a new concept, and its development history can be traced back as early as the semantic web [4] in the early 1960s. In 2012, Google proposed the term of KG to optimize search. The emergence of KG makes the web more intelligent and more in line with human thinking patterns. At present, there are countless mature projects about KG. The representative ones are Yago [5], DBpedia [6], Freebase [7], etc. Yago is a knowledge system based on public resources such as Wikipedia, WordNet, GeoNames, etc., which contains a large number of entities and relationships. DBpedia is a KG based on Wikipedia, which was first released in 2007, and now contains 4 million entities and 150 million relationships [8], covering most of the content in Wikipedia. Freebase is a generalized domain KG created by Google. The LinkedUP Project [9] is a pioneer in data openness in the field of education, since 2012, LinkedUP has focused on the issue of open data in education, aiming to advocate the use and creation of available data and to recognize the importance of open data. Knewton in the US builds a KG of the intersection of academic concepts [10], linking different disciplines, different school years, different knowledge points, and all learning resources together to form a vast KG, which enables the recommendation of personalized learning resources and learning paths.

Domestic better-developed KGs include zhishi.me [11], which is constructed based on several network resources such as Baidu Encyclopedia and Interactive Encyclopedia, CN-DBpedia [12], a KG based on Wikipedia, and SmartKG [13], which Microsoft develops. In the research of educational KG, Tsinghua University and Microsoft jointly released Open Academic Graph [14], which contains papers, authors, publication locations, and matching relationships between papers; Xu et al. proposed a collaborative filtering algorithm based on KG and constructed a K12 online education platform using the algorithm [15], which effectively solved the resource overload problem encountered in the process of students' online learning. The research team of Huazhong University of Science and Technology carried out research on the construction of a discipline KG based on latent semantic analysis.

The following conclusions can be drawn: First, the development of KG in the field of education has been rapidly changing in recent years; Second, there are more and more researchers and constructions related to the curriculum KG and they are widely used in teaching; Third, at present, there are fewer types of research on the KG of the secondary school curriculum in our country, which has an excellent space for development.

3   Design and Construction of Python KG

When constructing a KG, there are two construction ideas: top-down and bottom-up. The top-down construct refers to defining high-level knowledge first, and then defining low-level knowledge to form a complete knowledge system. Bottom-up refers to the process of building from the bottom entities and relationships first [16]. We use the bottom-up approach, and the specific steps include: collecting the underlying data through crawler technology, extracting entities and attributes from the collected data, extracting relationships between entities, and storing entities and relationships.

3.1   Acquisition of Data

The sources for acquiring Python course knowledge data include two aspects. One: textbooks, teaching aids and other books; and two: Python-related data in official Python documents and rookie tutorial websites [17].

Selenium is an open-source tool for web application testing; we chose to use Selenium to crawl the data because of the following advantages: it can control the browser can, avoid dynamic loading brought about by the incomplete page problem; simulate the user's operation, reduce the chances of being monitored by the anti-crawl [18].

Once you have decided to use Selenium to access a web page, you must choose how to get the information. The HTML in the source code of a web page can be seen as a structured text that contains many tags and links as well as textual information. Usually, regular expressions or Xpath can be used to process these texts. Xpath can be used to process the webpage information according to the tag name or attribute, which is more convenient than regular expressions [19], and the information processed by Xpath is the data needed to construct the Python KG for middle school. We choose to save the data as JSON files because the Python KG requires a small amount of data, and the JSON files can retain the structured information and are easy to read. The official Python tutorial page is shown in Fig. 1, from which it can be seen that the tutorial home page has a directory hierarchy. Figure 2 shows the structure of the second-level pages in the first-level headings, including a detailed description of the relevant first-level headings and related subheadings (a total of four levels of titles); these second-level pages of the hierarchy between the content and the headings is precisely

Fig. 1. Home page of the crawler page.   **Fig. 2.** Secondary page.

the construction of the KG of the data needed. Based on the catalog of the home page, we gradually obtain the content of each title and sub-title link, and save the obtained content according to the hierarchical structure of the catalog in the corresponding hierarchical structure, so that we can obtain a hierarchical structure of Python knowledge information. Finally, this data is saved in JSON format, and some of the data obtained by the crawler is shown in Fig. 3

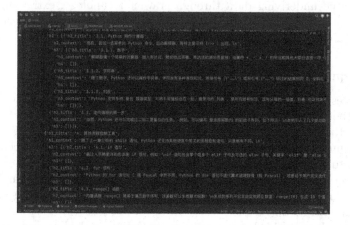

Fig. 3. Collected data.

3.2   Knowledge Extraction

Knowledge extraction methods include rule-based, machine learning-based, machine learning-based, and hybrid knowledge extraction based on machine learning and rules [20]. In this study, to ensure the accuracy of knowledge for the data obtained from web crawlers based on the THU Lexical Analyzer for Chinese (referred to as THULAC) model and manual calibration joint way for knowledge extraction, for textbook data using manual screening for knowledge extraction, relationships and attributes are obtained through manual screening.

Knowledge Extraction Based on the THULAC Model. The Python data we obtained through the crawler cannot provide a large amount of training data; usually speaking, the training of a machine learning entity extraction scheme needs to provide thousands of training data, while the total number of knowledge in the Python data is only one thousand. Therefore, we preprocess the crawled statements based on the THULAC model, and then filter them manually. In this way, the workload of manual screening work is reduced, and the accuracy of knowledge extraction is also guaranteed.

The crawled data are labeled with the THULAC and processed step by step according to the labeling level in the JSON data, and the results are shown in Fig. 4, where n is a noun, x is other, and v is a verb. As shown in the figure, the THULAC model classifies the word "Python" into other classes, and "Python" is indispensable knowledge in the actual construction of the KG, so it is necessary to manually calibrate the segmentation results to ensure the reliability of the KG. Therefore, it is necessary to calibrate the result manually to ensure the reliability of the KG.

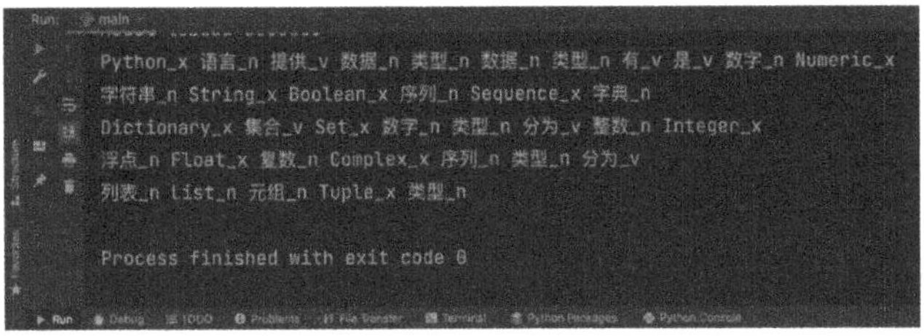

Fig. 4. THULAC segmentation result.

It is first extracted as entities, where nouns will be manually processed according to the needs of the KG in this paper, and specialized nouns or gerunds in Python will be saved as entities. Because attributes and entities have the property of adjacency, this work is done together in extracting entities.

The Python course knowledge entities have been identified through entity extraction, but the associative relationships between them remain unexplored, and the knowledge is fragmented. Therefore, the relationships between knowledge entities also need to be extracted to complete the construction of secondary Python KG. Relationship extraction is to identify the entities and their semantic relationships in the course knowledge data and represent the obtained course knowledge entities and the semantic relationships between entities in the form of "entity-relationship-entity" ternary, which facilitates the construction of subsequent KG.

Knowledge Extraction Based on Manual Screening. For the paper version of the textbook, if you use the same way as the crawler data for knowledge extraction, you need to scan the paper book as a picture first, then use the image recognition technology to transform the picture content into computer-recognizable text. In this process it is easy to produce errors such as data serialization and omission, which increases the workload of data cleaning in the later stage and affects the quality of knowledge, affecting the quality of the KG. To ensure the accuracy and reliability of the course KG, manual screening is chosen to extract the knowledge from the paper version of the textbook, and the process is as follows: First, read the entire book carefully and grasp the whole book; Second, focus on the table of contents and organize the entities in the table of contents, such as chapters and sections, and the relationships between them; Third, read the text carefully and manually organize the text in the material into entities, relationships, and attributes, and save them in the form of triples for subsequent storage. After this, the relationships between entities are obtained using the same approach as for the crawler data. To facilitate the subsequent storage of the knowledge, the extracted entity attributes and relationships are saved in CSV format as shown in Fig. 5 and Fig. 6, respectively.

Fig. 5. Entity-Attribute Storage.   **Fig. 6.** Relational Storage.

Knowledge Storage. We choose to use the Neo4j graph database to store the entities and relationships of the Python KG for middle school.Neo4j graph database can process data quickly and efficiently, and use its own Cypher query language to realize the addition, deletion, and modification of the Python KG.

Knowledge Visualization. Finally, we choose to use the Interactive Graph-neo4j tool to establish a connection between Interactive Graph and the Neo4j graph database to visualize the nodes and relationships stored in the Neo4j graph database, and to complete the construction of a mid-level Python KG.

The mid-level Python KG has three functions: the Knowledge Navigator, the Knowledge Browser, and the Relationship Finder. The knowledge navigator supports viewing all the knowledge in the entire KG and the relationships between knowledge and knowledge, as shown in Fig. 7. Clicking on a knowledge point allows you to view the attributes stored under this knowledge point, i.e., the descriptive information of the knowledge. The Knowledge Navigator also includes a variety of appearance settings such as displaying navigation buttons and adjusting theme colors.

The Knowledge Browser supports viewing information about a knowledge point and its associated knowledge. To find the knowledge involved in Fig. 8, for example, enter "Python control structure" in the search box, the interface will display the knowledge point in the form of a node, double-click on the knowledge node will display the number of knowledge nodes associated with the knowledge point and display the related knowledge around it: "loop structure", "sequential structure", "selection structure". If there is a next level of knowledge at this level, double clicking will further pop up the related knowledge points. In teaching,

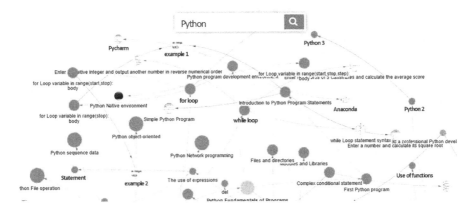

Fig. 7. Python KG.

teachers can use the application to show students the knowledge framework of the content learned in this lesson, so as to learn from the past and strengthen the connection between the old and new knowledge.

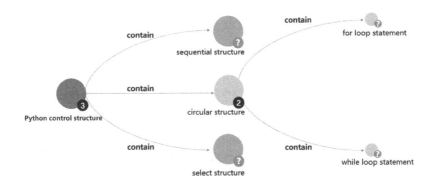

Fig. 8. Double-click node.

4 Implementation and Evaluation of Python KG

To verify its practical value, we applied Python KG as a teaching aid in a middle school in Jinan. We explored the effectiveness of middle school Python KG through a teaching experiment. The experiment was carried out in a secondary school in Jinan, and two classes of the second year of computer science majors were selected as the experimental subjects, of which class 1 was the practical class, which integrated Python KG in the teaching process; class 2 was the control class, which was taught by traditional teaching methods.

4.1   Analysis of Pre-test and Post-test Results

To ensure the reliability of the experimental results, the unit test results of the first chapter of Python Programming, fundamentals of Python Programming, were chosen as the pre-test data before conducting the teaching experiments. The test scores of class 1 and class 2 were statistically analyzed by SPSS data analysis software, and the results are shown in Table 1. According to the p-value in the Shapiro-Wilk test, it can be obtained that the scores of the two classes conform to normal distribution, the mean score and standard deviation of class 1 are 50.959 and 5.726, respectively. The mean score and standard deviation of class 2 are 51.958 and 5.787, respectively, and the mean scores and standard deviation of the two classes are relatively close to each other, so that the two classes are at similar levels of achievement.

Table 1. Results of Pre-test Performance Analysis

Name	Sample size	Average value	Standard deviation	Kolmogorov-Smirnov test		Shapiro- Wilk test	
				Statistic D value	p	Statistic W value	p
Class 1	49	50.959	5.726	0.148	0.009**	0.943	0.019*
Class 2	48	51.958	5.787	0.143	0.016*	0.946	0.027*

After completing the teaching of the second chapter, we conducted a post-test for class 1 and class 2 on the knowledge of Python program statements in the second chapter. The results of the analyzed post-test scores are shown in Table 2. According to the data in the table, it is known that the mean and standard deviation of the scores of the classes that have applied the Python KG are slightly higher than the mean and standard deviation of the classes that have not applied the Python KG. The results show that using Python KG in teaching can improve academic performance and proves the effectiveness of Python KG.

Table 2. Results of Posttest Performance Analysis

Name	Sample size	Minimum value	Maximum value	Averrage value	Standard deviation	t
Class 1	49	53	81	65.653	6,676	68.843
Class 2	48	48	75	57.896	5.969	67.201

4.2 Questionnaire Analysis

Based on the above teaching practice, this section chooses the questionnaire method to collect the trial data from the experimental group class students, and analyze the data to test the application effect of Python KG in secondary school.

The questionnaire survey used the overall sampling method to distribute the paper questionnaire to 49 students in the class, including 20 male students and 29 female students. It recovered the questionnaires when the students finished filling out the questionnaires, and the recovery rate of the questionnaires was 100%. First of all, the data were cleaned, eliminating invalid, duplicate, and irregular data in the questionnaire, such as incomplete answers or not by the requirements of the solutions, to obtain a valid questionnaire 48. Then the valid data were organized into a table form to facilitate the subsequent data analysis.

For the questionnaire survey, reliability and validity analysis are needed after recovering the questionnaire. The purpose of the reliability analysis of the questionnaire is to evaluate the stability and consistency of the questionnaire. We use SPSS statistical analysis software Cronbach's Alpha coefficient for the reliability analysis. The results are shown in Table 3, the Cronbach's Alpha value is 0.856, which indicates that the questionnaire has a good level of reliability.

Table 3. Confidence Analysis

Reliability statistics	
Cronbach's Alpha	*item count*
.856	16

Validity analysis refers to the assessment of the validity of the questionnaire, the results of the validity analysis are shown in Table 4, the Sig value is 0, which is less than 0.05, and the KMO (Kaiser-Meyer-Olkin) coefficient is 0.825, which is higher than 0.8, indicating that the questionnaire has good validity.

Table 4. Effectiveness Analysis

KMO and Bratlett validation		
KMO measure of sampling adequeacy		0.825
		1550.044
Bratlett's Test of sphericity	*Df*	180
	Sig.	0.000

Level of Satisfaction. Questions 1–4 of the questionnaire are specific to the satisfaction level of the system design, and this part mainly investigates four perspectives: interface, interactivity, operability, and application of the mid-career

Python KG. Question 1 targets the KG interface, mainly analyzing whether the interface is simple and straightforward; question 2 is to understand whether the interactivity of the KG is friendly, question 3 is to examine whether the operation of the KG is easy, and question 4 is to know whether the application designed based on the KG is reasonable and complete. The survey results are shown in Fig. 9, it has been shown that for the system design section students hold a high level of satisfaction with Python KG and only a few students hold a disagreement. Comparing the four questions, it is found that in the aspect of application design, the number of people who hold uncertain, disagree and strongly disagree is 8 %, which is slightly higher than the other three aspects.

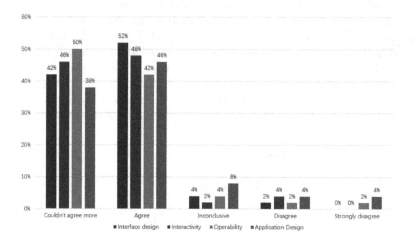

Fig. 9. System design attitudes.

Learning Support. Questions 5–9 are specific to the learning support dimension. Among them, question 5 is the knowledge presentation part; in order to verify whether Python KG can clearly display knowledge and the relationship between knowledge and knowledge, the results of the survey are shown in Figure Fig. 10, from which it can be seen that 94% of the students think that KG can clearly and systematically display the knowledge system of the Python course, which confirms that the KG has a powerful visualization function and that applying KG in teaching can innovate the way of presenting teaching content.

Questions 6–8 were specific questions about knowledge construction, and the detailed data are shown in Fig. 11. It can be seen that overall, the sum of the percentages of strongly agree and agree was more than 75 for questions 6, 7, and 8, and the sum of the percentages was the highest for question 8, which was more than 80%. The data show that more than 75% of the students believe that using KG for learning can help them better understand Python knowledge, which is conducive to linking previously learned knowledge with new knowledge, and more than 80% of the students believe that learning with the help of KG

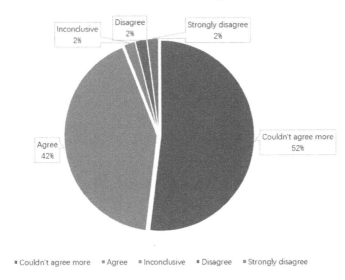

Fig. 10. Knowledge presentation attitudes.

can effectively sort out and summarize what they have learned, thus facilitating the meaningful construction of knowledge.

Learning Attitudes. Questions 9–12 are specific to the learning attitude dimension, including cognitive component, affective experience and behavioral intention. The specific data of cognitive component and affective experience are shown in Fig. 12. For the question of cognitive component, 73% of students think that using KG for learning is an effective way of learning. In terms of affective experience, more than 77% of students strongly agreed and agreed, i.e., 70% of students had a positive affective experience after using KG for learning, indicating that they were willing to spend more effort to learn Python.

Questions 11 and 12 are specific regarding behavioral intention, mainly exploring the impact that KG has on students' behavior, and the detailed data are shown in Fig. 13. The percentage of people with a disagreeing attitude is 6 %, and the number of people with a strongly disagreeing attitude is 0. The data indicate that students will actively choose to use KG as an aid to help them learn, and have a solid expectation for applying KG as a presentation method to other subjects.

The data from the questionnaire and interview method were summarized and the following conclusions were drawn:

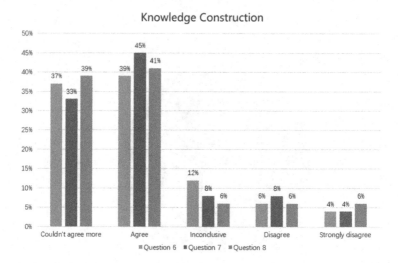

Fig. 11. Knowledge construction attitudes.

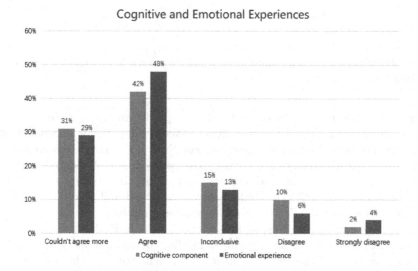

Fig. 12. Cognitive and emotional experience.

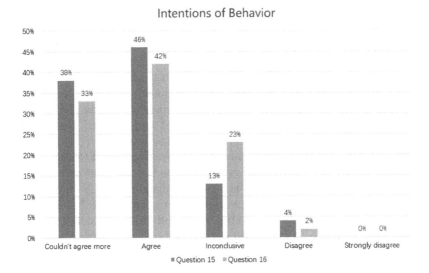

Fig. 13. Behavioral intentions attitudes.

– Python KG has a simple interface, good interactivity, and easy operation, and students can use it directly without training before applying it to teaching.
– Python KG is presented in a novel way, which can effectively improve students' learning interests.
– Python KG provides an optional knowledge framework that effectively reduces the knowledge load, helps students sort out difficult and essential knowledge, and promotes the meaningful construction of knowledge.

Our expected outcomes of constructing the Python KG have also been achieved: the Python KG could engage students by presenting Python concepts appealingly and interactively, increasing their interest in learning. Students are able to retain Python knowledge more effectively due to the visual and interactive nature of the KG, leading to improved performance in Python courses. By offering a comprehensive and structured repository, the Python KG helps overcome the challenge of information overload, making Python learning more accessible.

5   Limitations

This study involves the following limitations, which may be considered in future research studies: First, due to the small amount of course data involved in this study, the knowledge extraction phase was mainly based on manual intervention, which ensured quality as much as possible, but was less efficient.

Second, the teaching time in this study was oriented to a small number of subjects, and increasing the number of subjects taught would make the practice results more generalizable.

6   Conclusion

In this study, given the current students' learning needs and the problems existing in the teaching of Python courses, we have carried out a teaching design based on the Python KG, and applied teaching cases to teaching practice. At last, we collected data from the experimental group by questionnaire, tested the teaching effect of Python KG, and proved the effectiveness of applying KG to Python teaching from the dimensions of learning support and knowledge construction. Through teaching design and specific practice, this study reflected that the structure of curriculum KG can not only innovate the presentation of teaching content, but also provide the knowledge framework function to reduce the knowledge load. It can also promote students to construct knowledge, improve learning efficiency and quality meaningfully. In summary, constructing the Python KG can enhance the learning experience for Python students by offering a visually intuitive and comprehensive resource. It could address the issues of information overload and passive learning by making Python more accessible, engaging, and personalized.

Acknowledgement. This work is supported by Shandong Provincial Project of Graduate Education Quality Improvement (No. SDYJG21104, No. SDYJG19171), the Key R&D Program of Shandong Province, China (NO. 2021SFGC0104, NO. 20 21CXGC010506), the Natural Science Foundation of Shandong Province, China (No. ZR2020LZH008, ZR2021MF118, ZR2022LZH003), the National Natural Science Foundation of China under Grant (NO. 62101311, No. 62072290), the Postgraduate Quality Education and Teaching Resources Project of Shandong Province (SDYKC2022053, SDYAL2022060), the Shandong Normal University Research Project of Education and Teaching (No. 2019XM48), and Industry-University Cooperation and Education Project of Ministry of Education (No. 220602695231855).

References

1. Xu Jingcheng. Ministry of Education Issues Education Informatization 2.0 Action Plan[J]. Information Technology Education in Primary and Secondary Schools,2018,197(05):4
2. Ye Yuting. Smart education will break through the boundaries of schooling[N]. China Youth Daily,2023-02-14(002)
3. Tao W, Xiangwei Z, Lifeng Z, et al. A graph-based interpretability method for deep neural networks[J]. Neurocomputing,2023,555
4. Gu Yile. Patent analysis of knowledge mapping technology based on patent information measurement[J]. Journal of Library and Intelligence, 2021,6(07):45-52+61
5. Suchanek F M, Kasneci G, Weikum G. Yago: a core of semantic knowledge[C]. Proceedings of the 16th International Conference on World Wide Web. 2007: 697-706
6. Bizer, C., et al.: DBpedia - A crystallization point for the Web of Data[J]. Web Semantics: Science, Services and Agents on the World Wide Web **7**(3), 154–165 (2009)
7. Bollacker K, Evans C, Paritosh P, et al. Freebase: a collaboratively created graph database for structuring human knowledge[C]. Proceedings of the 2008 ACM SIGMOD International Conference on Management of Data. 2008: 1247-1250

8. Auer S, Bizer C, Kobilarov G, et al. Dbpedia: a nucleus for a web of open data[C]. The Semantic Web: 6th International Semantic Web Conference, 2nd Asian Semantic Web Conference, ISWC 2007+ ASWC 2007, Busan, Korea, November 11-15, 2007. Proceedings. Springer Berlin Heidelberg, 2007: 722-735
9. Herder E, Dietze S, D'Aquin M. LinkedUp-Linking Web data for adaptive education.[C]. UMAP Workshops. 2013
10. Dong Xiaoxiao, G., Hengnian, Z.D.: New research progress in knowledge graph and its application challenges in education[J]. Digital Education 8(05), 10–17 (2022)
11. Niu, X., Sun, X., Wang, H., Zhishi. me-weaving chinese linking open data[C]. The Semantic Web-ISWC, et al.: 10th International Semantic Web Conference, Bonn, Germany, October 23–27, 2011, Proceedings, Part II 10. Springer, Berlin Heidelberg **2011**, 205–220 (2011)
12. Xu B, Xu Y, Liang J, et al. CN-DBpedia: A never-ending Chinese knowledge extraction system[C]. Advances in Artificial Intelligence: from Theory to Practice: 30th International Conference on Industrial Engineering and Other Applications of Applied Intelligent Systems, IEA/AIE 2017, Arras, France, June 27-30, 2017, Proceedings, Part II. Cham: Springer International Publishing, 2017: 428 -438
13. Amr A, Garcia J D F, Maribel A, et al. SMART-KG: Hybrid Shipping for SPARQL Querying on the Web[J]. 2020
14. Huang, H., Wang, H., Wang, X.: An analysis framework of research frontiers based on the large-scale open academic graph[J]. Proceedings of the Association for Information Science and Technology **57**(1), e307 (2020)
15. Yajun, X., Jian, G.: Personalized learning resources recommendation for K12 learning platform[J]. Computer System Applications **29**(07), 217–221 (2020)
16. Jiarui, L., Huayu, L., Yang, Y., et al.: Construction and application of knowledge mapping for discipline construction based on event extraction[J]. Computer System Applications **31**(11), 100–110 (2022)
17. Yu, X.M.: Software and educational information service. Shandong People's Publishing House **2022**, 05 (2022)
18. Jiao, X., Yu, X., Peng, H., Zhang, X.: A Smart Learning Assistant to Promote Learning Outcomes in a Programming Course. International Journal of Software Science and Computational Intelligence (IJSSCI) **14**(1), 1–23 (2022)
19. Xuanchi C, Xiangwei Z, Kai S, et al. Self-supervised vision transformer-based few-shot learning for facial expression recognition[J]. Information Sciences,2023,634
20. Yongqiang Y, Xiangwei Z,Bin H, et al. EEG emotion recognition using fusion model of graph convolutional neural networks and LSTM[J]. Applied Soft Computing Journal,2021,100
21. Gong, Z., Yu, X., Fu, W., Che, X., Mao, Q., Zheng, X. (2021). The Construction of Knowledge Graph for Personalized Online Teaching. In: Tan, Y., Shi, Y., Zomaya, A., Yan, H., Cai, J. (eds) Data Mining and Big Data. DMBD 2021. Communications in Computer and Information Science, vol 1454. Springer, Singapore. https://doi.org/10.1007/978-981-16-7502-7-11

An Improved Prototypical Network for Endoscopic Grading of Intestinal Metaplasia

Rui Li, Xiaomei Yu$^{(\boxtimes)}$, Xuanchi Chen, and Xiangwei Zheng

School of Information Science and Engineering, Shandong Normal University, Jinan 250358,
China
yxm0708@126.com

Abstract. Intestinal metaplasia (IM) is confirmed to be the commonest symptom of early gastric cancer. IM grading by endoscopic images is essential to reduce the risk and mortality of gastric cancer. However, expensive and time-consuming manual annotations result in insufficient endoscopic image datasets to support the training of deep learning models. Prototypical network is the foundational model in few-shot learning and routinely is used to overcome insufficient datasets. Therefore, we propose an improved prototypical network (PN-ViT) for endoscopic grading of intestinal metaplasia. Firstly, the original IM dataset is divided into support sets and query sets, where the support set images are used to calculate prototypes and the images are only used to extract features in the query set. Second, a pre-trained vision transformer (ViT) is used as an embedding layer to extract IM lesion features from support sets. Then, the average value of features is calculated as the prototype of each category. Third, the features of the query set images are extracted using the pre-trained ViT as well. After the distances between them and prototypes are calculated for predicting the category. Finally, we conducted extensive experiments on a private intestinal metaplasia grading dataset from a Grade-A tertiary hospital. Experimental results showed that PN-ViT achieved the top classification accuracy of 70.2% in 3-way 5-shot scenario and verified the effectiveness of PN-ViT.

Keywords: Prototypical network · Few-shot learning · Endoscopic grading of intestinal metaplasia · Vision transformer

1 Introduction

The Gastric cancer (GC) is one of the most common lethal tumors worldwide, and patients with advanced GC have a substantial mortality rate, due to the fact that GC is quite often detected at an advanced stage and is extremely difficult to treat. While the survival rate of early GC is on the high side, it is often ignored or misdiagnosed due to atypical symptoms, resulting in missing the optimal time for treatment. Among the early symptoms arising from gastric cancer, intestinal metaplasia (IM) has been shown to be associated with the development of gastric cancer [4, 5]. GC typically progresses gradually from chronic gastritis, and IM is considered to be an intermediate stage of chronic gastritis, as well as an immeasurably important stage in the development of GC. Thus, individuals with IM in the gastric mucosa have a elevated risk of cancer, and the greater the degree of IM, the higher the risk of GC development. Given the association between IM and GC, the detection and monitoring of IM is of importance for high-risk patients, especially those with other GC risk factors. Early detection and intervention of IM can help prevent and control the progression of GC. Therefore, accurate IM grading is crucial for the detection of early gastric cancer [6].

IM can only be graded by specialized physicians. However, physicians usually identify lesions by visual observation, allowing fine-grained features to be ignored. In addition, there are subtle differences for physicians with different experience and knowledge in the IM grading. This leads to insufficient IM datasets with accurate labels and unbalanced distribution of categories. Deep learning techniques [2, 3] have received considerable attention in the field of medical image analysis and pathology as a powerful machine learning method based on artificial neural networks. Its superior pattern recognition and feature extraction capabilities have led to significant advances in tumor classification, disease prediction, and image analysis. But it is also notorious for requiring large datasets. Insufficient IM datasets are not suitable for training with deep learning. Furthermore, the IM grading is based on the lesion gastric parietal cells. As shown in Fig. 1, on the endoscopic images the lesion features of IM contain three aspects, namely bright blue crest, white nodules and the villus structure. These are elusive to distinguish and are poorly extracted by ordinary convolutional networks [7].

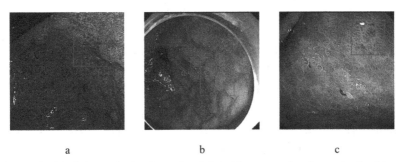

a   b   c

Fig. 1. (a) The labeled area in the figure is a bright blue crest, showing a slender, blue-white glandular structure. (b) The labeled area is a white nodule, which appears white with raised areas. (c) The labeled area in the figure is a villous-like feature

Learning from a few examples is the core issue of few-shot learning, which has achieved excellent performance in the medical imaging field. Jiang et al. [8] proposed a few-shot learning method for classification, which used transfer learning and meta-learning to solve sparse medical images. Sufficient experiments evidenced that the method achieved excellent results. Dai et al. [9] used a dual encoder structure to extract general and specific features of medical images to obtain enhanced features. Experimentally results proved that the method was 2.63% more accurate compared to the state-of-the-art method. In addition, Vision Transformer (ViT), a neural network model based on the attention mechanism, has been extremely successful in image classification. ViT is suitable for extracting endoscopic image features. First of all, endoscopic images usually have high resolution and complex structures. ViT can adaptively process images with different resolutions by modeling the image by dividing it into small blocks, capturing local features, and extracting tiny details, such as tissue structures and lesion areas [1]. Second, details are crucial for endoscopic images, and ViT utilizes a self- attention mechanism to interact and integrate the relationships of different regions during feature extraction, capturing subtle associations and extracting key features. In addition, endoscopic images are often affected by problems such as uneven illumination, noise and distortion. ViT is pre-trained to learn generic image feature representation with robustness and generalization capability. It can be finetuned based on the pre-trained model to adapt to various types of endoscopic images and extract high-quality and accurate feature. Yujin et al. [10] presented a hybrid 2-level visual converter for artificial intelligence-assisted 5-level pathologic diagnosis of gastric endoscopic biopsies, achieving excellent results and reducing the workload of limited pathologists.

Therefore, we propose an improved prototypical network (PN-ViT) for endoscopic grading of IM. First, the original IM dataset is divided into support sets and query sets. The support set is used to extract features and calculate prototypes, and the query set is available to validate the legitimacy of the prototype. Second, the vision transformer (ViT) is pre-trained on a large dataset. This is used as an embedding layer to extract IM lesion features from support sets. Third, according to category the average of the support set features is calculated as the prototype for each category. Fourth, the pre-trained ViT is adopted to extract the features of the query image. The distances between it and the prototypes are calculated to predict the category.

Our contributions can be summarized as follows:

(1) We apply few-shot learning to endoscopic grading of IM under insufficient endoscopic images, which can improve the efficiency of treatment.
(2) We propose an improved prototypical network (PN-ViT) for endoscopic grading of IM, in which ViT is used to extract fine-grained features in the IM images and distinguish diseased features.
(3) We conducted extensive experiments on a private IM dataset from a Grade-A tertiary hospital. The results show that PN-ViT has superior performance, achieving an accuracy of 70.2% on 3-way 5-shot.

The rest of the paper is structured as follows. Section 2 reviews related work. Our model (PN-ViT) and the implementation process of each module are introduced in Sect. 3. Section 4 designs experiments and discusses the experimental results. Section 5 describes the comprehensive conclusions.

2   Related Work

2.1   Endoscopic Grading of Intestinal Metaplasia

Intestinal metaplasia (IM) is a common stomach disorder that usually presents in the context of chronic gastritis, Helicobacter pylori infection and gastric ulcers. IM is generally categorized into three different grades: Grade 0, Grade 1 and Grade 2. The IM grading can assist physicians in diagnosing a patient's condition and designing a treatment strategy. Early automatic grading researches mainly relied on manual feature extraction methods. Diseased features are obtained by feature extraction methods and then they are graded using a basic classifier, such as a support vector machine (SVM). For example, Van et al. [11] proposed a method to extract pathological features and colors from endoscopic images, and classify them using SVM, which achieved an accuracy of 75%. Kanesaka et al. [12] proposed a classification method. This first extracts the eighth gray-level co-occurrence matrix (GLCM) features of the images, after which the coefficient of variation of eight GLCM features is calculated. Finally, the classification is implemented using SVM. Although manual feature extraction methods can accurately detect the condition of the gastrointestinal tract, they frequently suffer from overfitting and cannot automatically detect the nuances of endoscopic images [13, 14]. In modern years, the rapid development of deep learning technology [15] has been widely used in various fields, especially convolutional neural networks have made considerable achievements in the medical imaging field. Xu et al. [16] constructed a deep convolutional neural network system called Endangel to detect gastric precancerous lesions by image-enhanced endoscopy (IEE). It achieved excellent performance in IM grading. Lin et al. [17] improved the TResNet and achieved high accuracy in IM diagnosis. While deep learning algorithms have achieved superior performance in IM grading, they require sufficient datasets. And sufficient datasets are inaccessible to us, due to patient privacy and intractable labeling. Accordingly, an improved method for endoscopic grading of IM using few-shot learning is proposed to us.

2.2 Prototypical Network

The prototypical network was proposed by Snell et al. [18] in 2017. It considered the existence of a category induction for each category in the vector space. The average of sample vectors within a category is calculated as the prototype of that category. Then the distance between the prototype and the query sample is calculated for each category to achieve the classification. It has achieved a remarkable performance in image classification and is routinely used as an improved basic network by researchers. Some have investigated how to improve prototype calculation modules. For instances, Mishra et al. [19] enhanced the prototype network utilizing an attention mechanism. This is adaptively able to assign distinct weights to each sample in order to provide a more typical prototype of a category. Liu et al. [20] in 2020 proposed a bias reduction method for prototype correction, by which both intra-class and cross-class biases can be reduced. In addition, there are a host of researchers on how to optimize the feature extraction algorithm. Such as Tang et al. [21] proposed a multiscale spatial-spectral feature extraction algorithm based on the ladder structure, which was used to achieve the classification of hyperspectral images. Zhang et al. [22] used cosine margin to discriminate the acquired features to solve data bias. It achieved promising results in remote sensing scene classification. Although they achieved favorable performance, these models were proposed for specific scenarios. We are incapable of applying them to the IM grading. To overcome the challenge of extracting fine grain features from gastric parietal cells, endoscopic image features are extracted by large-scale network ViT. Then the prototype is calculated for prediction.

2.3 Vision Transformer

In modern times, vision transformers (ViT) have become a hot research topic in computer vision, which is a deep neural network based on a self-attention mechanism [23]. It segments images into small patches and converts these patches into sequential data for processing. The original version of ViT was proposed by Dosovitskiy et al. [24] in 2020. An encoder containing multiple transformer blocks was used. Each of these transformer blocks contains a multi-headed self-attentive mechanism and a feedforward neural network, as well as a few residual connections and layer normalization. Since its release, ViT has been used in medical imaging by numerous researchers. For case in point, Sharif et al. [25] combined ViT and CNN in a parallel manner, enabling them to extract features from images separately. And experiments were conducted on 10 datasets, achieving state-of-the-art results. Huang et al. [26] fused ViT with attention mechanism integrated convolution (AMC) blocks, due to the high interpretability of the ViT-based deep learning module. Experimental results showed that the model had excellent generalization ability and surpasses state-of-the-art methods. In conclusion, ViT has become an exciting technology in medical imaging. It has shown promising performance in a variety of tasks and is constantly being enhanced and optimized. Therefore, ViT is introduced into few-shot learning. It acts as a feature extractor to extract the fine-grained features of the image and achieve IM grading.

3 The Proposed Method

3.1 Overview

PN-ViT adopt an N-way K-shot meta-learning strategy, where N denotes the number of categories within an episode and K denotes the number of samples contained in each category in the support set. Specifically, the dataset $D = \{(x_1, y_1), (x_2, y_2), \ldots, (x_n, y_n)\}$ is divided into two parts, the training dataset D_{train} and the testing dataset D_{test} respectively. N categories are randomly selected from the training set D_{train}. K samples are drawn for these selected categories as the support set S. And samples are drawn from the remaining samples to obtain the query set Q. Given a support set, few-shot learning is minimizing the sample prediction loss in the query set. Overview of the PN-ViT is shown in Fig. 2. It is divided into two main modules: the feature extractor and the prototype calculation module. For each pair of support set and query set images (S, Q), we first extract the feature maps $S_{emb} = f_{emb}(S)$, $Q_{emb} = f_{emb}(Q)$ using ViT. S_{emb} and Q_{emb} represent the set of feature vectors for the support set and query set respectively. We calculate the average of the sample feature vectors of the support set as the prototype p_i for each category. Then the distance between Q_{emb} and the prototype p_i is calculated and used to classify the samples in the query set Q.

In the following, we first explain ViT for the feature extractor in Sect. 3.2. In Sect. 3.3, the computational procedure for computing the support set sample prototype is described.

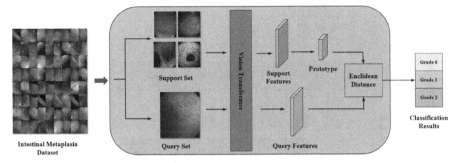

Fig. 2. Overview of the PN-ViT

3.2 ViT Feature Extractor

IM is graded according to the lesion degree of the gastric parietal cells. Thus, the fine-grained features of the gastric parietal cells were extracted by the attention-based ViT. The introduction of attention allows the model to focus more on the lesion region, which can improve the classification accuracy. The model is shown in Fig. 3, we reshape the image $x \in R^{H \times W \times C}$ into a patch sequence $x_p \in R^{N \times (P^2.C)}$, where here H and W are the height and width of the input image respectively. C is the number of channels. (p, p) denotes the height and width of each patch image. And $N = HW/P^2$ is the number of patches obtained, which is also the the length of the input sequence of transformer. The dimension of embedding we use throughout the transformer process is D. Therefore, we input the patch sequence x_p into the linear projection layer E to obtain the output (Eq. 1), which is called patch embedding.

$$z_0 = \left[x_{class}; x_p^1 E; x_p^2 E; \ldots; x_p^N E \right] + E_{POS}, E \in R^{(P^2.C) \times D}, E_{pos} \in R^{(N+1) \times D} \quad (1)$$

Similar to the [class] token of BERT, we add a learnable embedding ($z_0^0 = x_{class}$) before the patch embedding sequence in order to obtain the classification information. The size of the embedding obtained after adding the class token is $(N+1) \times D$. After that, a position code is added to each embedding to preserve the position information (Eq. 1). Where E_{POS} represents the position code information. The obtained embedding sequence is used as the input to the encoder, which is identical to the encoder in the original transformer [27]. It will not be explained in detail here. This consists of alternating combinations of multi-headed attention MSA and MLP blocks. LN layers are used before each block and residual connections are used after each block (Eqs. 2,3). At the end, we extract class token as embedding for prototype computation to achieve classification. Here, the image features are extracted by the ViT and we write the process as $z = f_{emb}(x_k)$.

$$z\prime_l = MSA(LN(z_{l-1})) + z_{l-1}, l = 1 \ldots L \quad (2)$$

$$z_l = MLP(LN(z\prime_l)) + z\prime_l, l = 1 \ldots L \quad (3)$$

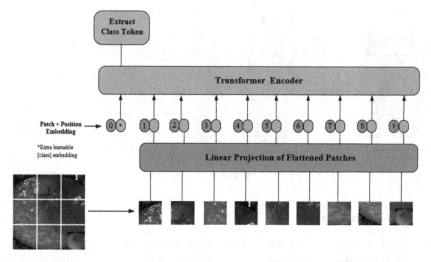

Fig. 3. ViT feature extractor

3.3 Prototype Calculation

After the above feature extraction module, the sets S_{emb} and Q_{emb} of the support set and query set are obtained respectively, where $S_{emb} = \{z_{1k}, z_{2k}, \ldots, z_{nk}\}$ and $Q_{emb} = \{z'_{1k}, z'_{2k}, \ldots, z'_{mk}\}$. The vectors inside S_{emb} and Q_{emb} set are all d-dimensional feature vectors. The average of the feature vectors in each class support set is calculated as the prototype of that category (Eq. 4).

$$p_k = \frac{1}{|S_{emb}^k|} \sum z_{ik} \tag{4}$$

where S_{emb}^k is the kth category in the support set S, and z_{ik} denotes the feature vector of the ith sample in the kth category of the support set. After that, the distances of all sample points $Q_k^{emb} = \{z'_{1k}, z'_{2k}, \ldots, z'_{mk}\}$ to each category prototype p_i are calculated in the query set Q. Finally, the *softmax* is used to predict the labels of the images. The category distribution of its query points is (Eq. 5):

$$p(y_k = c|z_{ik}, p_k) = softmax(-d(z_{ik}, p_k)) \tag{5}$$

where y_k denotes the label of the kth category in the support set.

4   Experiments and Discussion

4.1   Dataset and Setting

Dataset and Metrics. The IM Endoscopy dataset was obtained from a Grade-A tertiary hospital and covers images of patient visits from 2018 to 2021. Due to patient privacy protections and a lack of data, this hospital was unavailable to provide details such as the number and age of patients. Annotation of the dataset was conducted by four specialized physicians trained in the EGGIM scoring system. To ensure the quality of the dataset, we first performed a preliminary processing of the dataset. Eight graduate students (supervised by the professional endoscopist) were assembled to perform an initial screening of the dataset, eliminating images that were blurry or difficult to label. The screened images were manually labeled by two physicians with more than 5 years of experience in gastroscopy. During the annotation process, if there were discrepancies in the annotation results for the same image, we asked four experts, including these two experts, to reassess the extent of IM lesions in that image. If a consensus still could not be reached in the end, we removed this image from the dataset. Through this series of dataset processing, we obtained 1115 images as the final dataset, and the images were all in jpg format. The dataset contains 3 categories, classified as 0, 1 and 2 according to the severity of the lesions. Among them, Grade 2 indicates the most severe lesion, Grade 1 indicates a slight lesion, and Grade 0 indicates no lesion. Since IM dataset has fewer classes and cannot be divided according to the conventional few-shot learning method.

Table 1. Distribution details of the IM severity dataset

	Training set	Value set	Test set	Total data
Grade 0	247	93	88	428
Grade 1	238	74	74	386
Grade 2	184	56	61	301

As shown in Table 1, we divide three classes into training set, evaluation set, and test set according to a certain proportion. Where 60% of the training set contains 669 images, 20% of the evaluation set contains 223 images, and 20% of the test set contains 223 images. We use the accuracy to evaluate our model, acc calculated as follows (Eq. 6).

$$acc = \frac{right}{all} \tag{6}$$

where $right$ denotes the number of correctly predicted samples, all denotes the number of all samples.

Experimental Setup. We conduct experiments on the IM dataset. Since the dataset only includes 3 classes, we cannot use the conventional N-way, K-shot meta-learning method for training. We fix N as 3, and the samples selected from the training set are fixed 3 classes each time. The input images are resized 84×84, and the required patch size in ViT is set to 16×16. The Adma optimizer is used to optimize the model in the experiment. Two types of tasks are set for this dataset: 3-way 5-shot task and 3-way 1-shot task. We set the initial learning rate to 0.001 and reduce the learning rate by halves for every 20 epochs. The parameter patience is set to 50. The iteration is stopped when the best model parameters are still not updated after patience epochs. For each class of tasks, 200 epochs are trained, and 1000 epochs are iterated during testing. All experiments covered in this paper were run on a gpu with cuda version 11.6 and pytorch version 1.13.1+cu116.

4.2 Comparison with State-Of-The-Art

We compare PN-ViT with the model in Table 2. It shows the accuracy of the IM dataset on both the 3-way 5-shot and 3-way 1-shot tasks. As shown in Table 2, PN-ViT performed exceptionally well in both tasks. It improves the accuracy by more than 10% compared to most models on the IM dataset. Cov4-64 is used to extract image features in MN [28] and MAML [29]. Specifically, PN-ViT improves 12.5% and 15.6% in the 3-way 1-shot task and the 3-way 5-shot task in MN respectively. In MAML, 14.2% and 15.2% are improved in the 3-way 1-shot task and the 3-way 5-shot task respectively. GNN [30] uses graph neural networks to extract image features. Compared with it, PN-ViT improves 16.9% and 12.0% in the 3-way 1-shot task and the 3-way 5-shot task respectively. PN-ViT improved 4.9% and 3.4% in the 3-way 5-shot and 3-way 1-shot tasks, respectively, compared to the TPN [31] based on transductive inference. To demonstrate the superiority of ViT for extracting pathological features in the prototypical network, we compare the prototypical network with our model using Conv4, ResNet12 and ResNet50 as the feature extraction network, respectively, and the results show that PN-ViT has excellent performance. Among them, PN-ViT improved by 2.2% and 3.1% in the 3-way 1-shot task and the 3-way 1-shot task respectively compared to the original prototypical network. The superiority of the improved prototype network in IM endoscopic image classification is demonstrated.

Table 2. Comparison of classification accuracy of few-shot learning on IM dataset

Model	Backbone	3-way	
		1-shot	5-shot
MN [28]	Conv4	0.520	0.546
MAML [29]	Conv4	0.503	0.550
GNN [30]	GNN	0.476	0.582
PN [18]	Conv4	0.623	0.671
	ResNet12	0.631	0.678
	ResNet50	0.639	0.680
TPN [31]	Conv4	0.611	0.653
PN-ViT (ours)	ViT	**0.645**	**0.702**

4.3 Visualization

Figure 4 shows a visualization of the feature mapping. We show three pairs of support and query images and the query feature mappings for the above model and PN-ViT. It is found that (1) the results of two models, PN and PN-ViT, are significantly better than the other three models. Although MAML achieves better results in the third row relative to GNN and MN, it is still inferior to the results of PN-ViT. This indicates that GNN, MN and MAML have some problems in feature extraction, which leads to lower accuracy. (2) Compared with PN, there is not much difference in the results of PN-ViT. However, it can still be seen that the image features extracted by PN-ViT are closer to the lesion, which is the reason why PN-ViT is more accurate than PN.

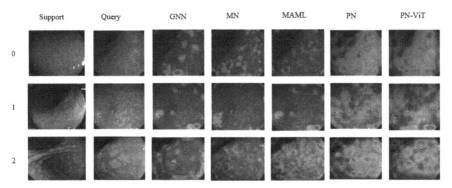

Fig. 4. Query feature maps from different methods for 5-way 1-shot tasks on dataset of IM

5   Conclusion and Future Work

To overcome the challenging task of training deep learning models due to insufficient effective labeled samples and difficulty extracting fine-grained features in IM dataset, we propose an improved prototypical network for endoscopic grading of intestinal metaplasia. Firstly, the original IM dataset is divided into support sets and query sets. Second, version transformer is pre-trained on a large dataset. This is used as an embedding layer to extract IM lesion features from support sets. Third, the average value of features is calculated as the prototype of each category. Then, the distances between the query image and prototypes are calculated for predicting the category. Finally, we conducted extensive experiments on a private intestinal metaplasia grading dataset from a Grade-A tertiary hospital. The substantial experimental results show that PN-ViT not only led to improved classification accuracy, but also extract fine-grained features more accurately on the IM dataset. PN-ViT achieves the top classification accuracy of 70.2% in 3-way 5-shot scenario, which proves the effectiveness of PN-ViT. In future work, we will consider assigning weights to each sample of the support set for calculating representative prototypes.

Acknowledgements. This work is supported by Shandong Provincial Project of Graduate Education Quality Improvement (No. SDYJG21104, No. SDYJG19171), the Key R&D Program of Shandong Province, China (NO. 2021SFGC0104, NO. 2021CXGC010506), the Natural Science Foundation of Shandong Province, China (No. ZR2020LZH008, ZR2021MF118, ZR2022LZH003), the National Natural Science Foundation of China under Grant (NO. 62101311, No. 62072290), the Postgraduate Quality Education and Teaching Resources Project of Shandong Province (SDYKC2022053, SDYAL2022060), the Shandong Normal University Research Project of Education and Teaching (No. 2019XM48), and Industry University Cooperation and Education Project of Ministry of Education (No. 220602695231855).

References

1. Yu, X.M.: Software and educational information service. Shandong People's Publishing House **2022**, 05 (2022a)
2. Shi, Y., Zheng, X., Zhang, M., Yan, X., Li, T., Yu, X.: A study of subliminal emotion classification based on entropy features. Front. Psychol. **13**, 781448 (2022). https://doi.org/10.3389/fpsyg.2022.781448
3. Jiao, X., Yu, X., Peng, H., Zhang, X.: A smart learning assistant to promote learning outcomes in a programming course. Int. J. Softw. Sci. Comput. Intell. (IJSSCI) **14**(1), 1–23 (2022)
4. Esposito, G., et al.: Endoscopic grading of gastric intestinal metaplasia (EGGIM): a multicenter validation study. Endoscopy **51**(06), 515–521 (2019)
5. Lee, J.W.J., et al.: Severity of gastric intestinal metaplasia predicts the risk of gastric cancer: a prospective multicentre cohort study (GCEP). Gut **71**(5), 854–863 (2022)
6. Marcos, P., et al.: Endoscopic grading of gastric intestinal metaplasia on risk assessment for early gastric neoplasia: can we replace histology assessment also in the West? Gut **69**(10), 1762–1768 (2020)
7. Wang, T., Zheng, X., Zhang, L., Cui, Z., Chunyan, Xu.: A graph-based interpretability method for deep neural networks. Neurocomputing **555**, 126651 (2023)
8. Jiang, H., Gao, M., Li, H., Jin, R., Miao, H., Liu, J.: Multi-learner based deep meta-learning for few-shot medical image classification. IEEE J. Biomed. Health Inform. **27**(1), 17–28 (2023)
9. Dai, Z., et al.: PFEMed: few-shot medical image classification using prior guided feature enhancement. Pattern Recogn. **134**, 109108 (2023)
10. Oh, Y., Bae, G.E., Kim, K., Yeo, M., Ye, J.C.: Multi-scale hybrid vision transformer for learning gastric histology: AI-based decision support system for gastric cancer treatment. IEEE J. Biomed. Health Inform. **27**, 4143 (2023)
11. van der Sommen, F., Zinger, S., Schoon, E.J., de With, P.H.N.: Supportive automatic annotation of early esophageal cancer using local gabor and color features. Neurocomputing **144**, 92 (2014)
12. Kanesaka, T., et al.: Computer-aided diagnosis for identifying and delineating early gastric cancers in magnifying narrow-band imaging. Gastrointest. Endosc. **87**, 1339 (2018)
13. de Souza, L.A., et al.: A survey on Barrett's esophagus analysis using machine learning. Comput. Biol. Med. **96**, 203 (2018)
14. Ali, H., Sharif, M., Yasmin, M., Rehmani, M.H., Riaz, F.: A survey of feature extraction and fusion of deep learning for detection of abnormalities in video endoscopy of gastrointestinal-tract. Artif. Intell. Rev. **53**, 2635 (2020)
15. Yin, Y., Zheng, X., Hu, B., Zhang, Y., Cui, X.: EEG emotion recognition using fusion model of graph convolutional neural networks and LSTM. Appl. Soft Comput. **100**, 106954 (2021)
16. Xu, M., et al.: Artificial intelligence in the diagnosis of gastric precancerous conditions by image-enhanced endoscopy: a multicenter, diagnostic study (with video). Gastrointest Endosc. **94**. 540 (2021)
17. Lin, N., et al.: Simultaneous recognition of atrophic gastritis and intestinal metaplasia on white light endoscopic images based on convolutional neural networks: a multicenter study. Clin. Transl. Gastroenterol. **12**, e00385 (2021)

18. Snell, J., Swersky, K., Zemel, R.: Prototypical networks for few-shot learning. In: Advances in Neural Information Processing ssystems 30 (2017)
19. Mishra, N., Rohaninejad, M., Chen, X., Abbeel, P.: A simple neural attentive meta-learner. arXiv preprint arXiv:1707.03141 (2017)
20. Liu, J., Song, L., Qin, Y.: Prototype rectification for few-shot learning. In: Vedaldi, A., Bischof, H., Brox, T., Frahm, J.-M. (eds.) ECCV 2020. LNCS, vol. 12346, pp. 741–756. Springer, Cham (2020). https://doi.org/10.1007/978-3-030-58452-8_43
21. Tang, H., Huang, Z., Li, Y., Zhang, L., Xie, W.: A multiscale spatial–spectral prototypical network for hyperspectral image few-shot classification. IEEE Geosci. Remote Sens. Lett. **19**, 1 (2022)
22. Zhang, X., Wei, T., Liu, W., Xie, Y.: Cosine margin prototypical networks for remote sensing scene classification. IEEE Geosci. Remote Sens. Lett. **19**, 1 (2022)
23. Chen, X., Zheng, X., Sun, K., Liu, W., Zhang, Y.: Self-supervised vision transformer-based few-shot learning for facial expression recognition. Inform. Sci. **634**, 206 (2023)
24. Dosovitskiy, A., et al.: An image is worth 16×16 words: transformers for image recognition at scale. arXiv preprint arXiv:2010.11929 (2020)
25. Sharif, M.H., Demidov, D., Hanif, A., Yaqub, M., Xu, M.: TransResNet: integrating the strengths of ViTs and CNNs for high resolution medical image segmentation via feature grafting (2022)
26. Huang, P., et al.: A ViT-AMC network with adaptive model fusion and multiobjective optimization for interpretable laryngeal tumor grading from histopathological images. IEEE Trans. Med. Imaging **42**, 15 (2022)
27. Vaswani, A., et al.: Attention is all you need. In: Advances in Neural Information Processing Systems 30 (2017)
28. Vinyals, O., Blundell, C., Lillicrap, T., Wierstra, D.: Matching networks for one shot learning. In: Advances in Neural Information Processing Systems 29 (2016)
29. Finn, C., Abbeel, P., Levine, S.: Model-Agnostic Meta-Learning for Fast Adaptation of Deep Networks. Cornell University Library, Ithaca (2017)
30. Garcia, V., Bruna, J.: Few-shot learning with graph neural networks. arXiv preprint arXiv: 1711.04043 (2017)
31. Liu, Y., et al.: Learning to propagate labels: transductive propagation network for few-shot learning. arXiv preprint arXiv:1805.10002 (2018)

Secure Position-Aware Graph Neural Networks for Session-Based Recommendation

Hongzhe Liu⬛, Fengyin Li(✉)⬛, and Huayu Cheng

School of Computer Science, Qufu Normal University, Rizhao, China
lfyin318@qfnu.edu.cn

Abstract. Session-based recommendation, a specific type of recommendation system, leverages users' interaction sequences to provide recommendations. Unfortunately, these approaches tend to overlook user privacy protection and are susceptible to session sequence leakage. By introducing BGV homomorphic encryption and position information into session-based recommendation, a secure position-aware session-based recommendation is proposed. We propose the secure session-based recommendation (SSBR) and position-aware graph neural network (PA-GNN). By leveraging item embedding learning and session embedding learning with graph neural network in both local and global contexts, based on BGV fully homomorphic encryption, we provide efficient and secure session recommendations for users. While ensuring user privacy preservation, our proposed model demonstrates superior performance over state-of-the-art baselines in session-based recommendations (SBRs), as indicated by the experimental results on two benchmark datasets.

Keywords: Secure · session-based recommendation · graph neural network

1 Introduction

As the demand for personalized recommendation systems continues to grow in modern society, the massive data generated by users on internet platforms has become the cornerstone of recommendation algorithms. However, traditional recommendation methods involve the collection, storage, and analysis of a large amount of users' personal information, which poses a risk of privacy leakage. Users' browsing history, purchase records, interests, and hobbies, among other sensitive data, serve as inputs to algorithms to achieve more personalized and accurate recommendation results.

The collection and storage of personal information are the main sources of privacy leakage. Recommendation systems need to obtain sensitive data such as users' identity, geographical location, preferences, and emotional inclinations to establish user profiles and behavioral models. Without adequate protection, this data is vulnerable to hacker attacks or misuse by internal personnel, resulting in the risk of user privacy exposure [1].

Data sharing and cross-analysis are another factor leading to privacy leakage. Partners and third-party service providers may share user data with recommendation systems for

J. Vaidya et al. (Eds.): AIS&P 2023, LNCS 14510, pp. 134–146, 2024.
https://doi.org/10.1007/978-981-99-9788-6_11

advertising and cross-recommendation purposes. However, this data sharing increases the risk of user data being correlated and subjected to cross-platform analysis, further exposing users' sensitive information.

Additionally, data storage and transmission also present potential privacy risks. User data is stored on recommendation system servers and transmitted over networks. Without appropriate encryption measures and security safeguards, it may be susceptible to hacker attacks or man-in-the-middle attacks, leading to user privacy breaches [2].

Therefore, traditional recommendation methods often overlook the importance of privacy protection for users, which can potentially result in privacy leakage during the transfer of user information. To address these concerns, we propose two models: secure session-based recommendation (SSBR) and position-aware graph neural network (PA-GNN). Our approach combines the use of homomorphic encryption and session recommendation techniques: For homomorphic encryption, we employ the efficient BGV scheme. For session-based recommendation, we utilize both global and local methods to learn item embeddings and session embeddings independently.

2 Secure Session-Based Recommendation

In this section, we present our secure session-based recommendation (SSBR) model. We employ the BGV homomorphic encryption scheme, which enables fully homomorphic encryption. This scheme operates by representing both plaintext and ciphertext as polynomials, allowing encryption and computation operations to be performed on the polynomial ring. By representing our data as polynomials, we can effectively utilize the BGV homomorphic encryption scheme.

The specific process of our homomorphic encryption scheme is as follows: As an e-commerce platform (client), we possess the session sequences of the user's interactive products. We upload this data to a local database for storage, and before providing the data to the cloud server, the database is homomorphically encrypted. Subsequently, the cloud server receives the encrypted session sequences and employs the homomorphic method to train them by the neural network. The cloud server then returns the predicted results in the form of ciphertext to the client. Finally, the client decrypts the received ciphertext to obtain the final predicted recommendation results, which are then used for making recommendations to the user. The specific process is illustrated in Fig. 1 [3, 4].

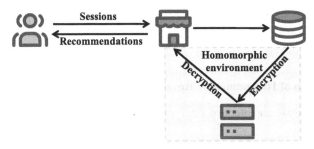

Fig. 1. Specific process of SSBR model.

2.1 Approximation of Activation Function

Our objective is to develop a graph neural network that operates within the limitations of homomorphic encryption. However, in the homomorphic encryption scheme, it is not possible to directly apply activation functions to ciphertexts. As a result, we need to identify an appropriate alternative scheme that enables arithmetic operations to be performed on ciphertexts.

Table 1. Use of operators in neural networks.

Operation	Homomorphic	Example
$+$	Yes	$\sum\limits_{v_i \in S} \mathbf{h}^S_{v_i,k}$
$-$	Yes	$e^x - e^{-x}$
$\times$	Yes	$\mathbf{s}_k \odot \mathbf{c}_{j,k} \|w_{ij}\| \mathbf{p}_\mu^{r_{i,j}^g}$
$\div$	No	$\frac{1}{1+e^{-x}}$
e^x	No	e^{-x}

As depicted in Table 1, in the five operations (addition, subtraction, multiplication, division, and exponentiation), the first three are homomorphic operations, whereas the last two are not supported in the BGV scheme [5]. These operations are essential for the activation function Sigmoid commonly used in neural networks. Consequently, the BGV encryption scheme does not provide support for performing these operations on ciphertexts. In order to facilitate secure computation of the Sigmoid function, we approximate it by substituting it with the Taylor expansion,

$$y = \frac{1}{1+e^{-x}} = \frac{1}{2} + \frac{x}{4} - \frac{x^3}{48} + o(x^4) \approx \frac{1}{2} + \frac{x}{4} - \frac{x^3}{48} \approx 0.5 + 0.25x - 0.02x^3 \quad (1)$$

By employing the Taylor expansion approximation, the estimation of the activation function involves only addition and multiplication operations. This effectively resolves the challenge of being unable to utilize division and exponentiation operations on BGV ciphertexts, which is consistent with other activation functions commonly used in neural networks [6].

2.2 Application of Homomorphic Encryption

Encryption of session sequences. Encrypting the session sequence is a prerequisite for cipher training.

ParamGen(λ, μ, b) $\rightarrow$ (*para*). Enter the security parameters λ and μ, and a bit b to get the system public parameter (*para*) = ($q, d, n, N, \mathcal{X}$).

PrivateKeyGen$(q, d, n, N, \mathcal{X}) \rightarrow sk$.Input the system parameter $q, d, n, N, \mathcal{X}$ to generate a vector t and each of its elements $t[i]$ obeys the $\mathcal{X}$ distribution sampling. The private key $sk \leftarrow (t[1], t[2], \cdots, t[N])[2], \cdots, t[N])$ is then obtained.

PubKeyGen$(q, d, n, N, \mathcal{X}, sk) \rightarrow pk$. Input the system parameters $q, d, n, N, \mathcal{X}$ and the private key sk, generate a random matrix B and a vector e, and each element of the vector $e[i]$ obeys the distribution sampling of $\mathcal{X}$. Multiply the matrix B with the vector t and add the vector $2e$ to get the vector b. The vector b is right concatenated with the matrix $-B$ to get the matrix A, which is used as the user's public key pk.

Encrypt$(m, pk, q, d, n, N, \mathcal{X}) \rightarrow c$. The input parameter m is the plaintext message, pk is the public key, the plaintext vector $m = (v_{s,1}, v_{s,2}, \ldots, v_{s,n})$ is known, a vector r is randomly generated and each of its elements $r[i]$ is randomly taken to be 0 or 1, the public key matrix A is transposed and multiplied by the vector r, the result obtained is added with the plaintext vector m and the resulting ciphertext $c = (v'_{s,1}, v'_{s,2}, \ldots, v'_{s,n})$ is obtained [7, 8].

Cipher Training. In this phase, a clever integration of BGV homomorphic encryption and session-based recommendation takes place.

Ciphertext Network. In the BGV scheme, the weights and biases of the neural network need to be converted into ciphertext form. This involves encrypting the weights and biases using a cryptographic algorithm and representing them in the form of vectors in the BGV scheme [5, 9].

Forward Propagation of Secret Message. Homomorphic multiplication operations are performed on the encrypted input features and encrypted weights to obtain the encrypted linear transformation results.

An approximate activation function is applied to the encrypted linear transformation result, and it is computed using a homomorphic encryption algorithm.

For multilayer neural networks, the ciphertext linear transformation and ciphertext activation function operations are repeated until the last layer is reached.

Back Propagation of Ciphertext. Decrypts the ciphertext output to plaintext form. The loss function is computed using the plaintext output and the labels to obtain the loss value in the form of the plaintext.

Loss values in plaintext form are encrypted using homomorphic encryption algorithms that convert loss values in plaintext form to ciphertext form.

The ciphertext backpropagation is performed layer by layer starting from the last layer. For each layer, the backpropagation computation is performed using a homomorphic encryption algorithm, which includes computing the gradient and weight update.

The computed gradient is applied to the weight update using homomorphic encryption algorithm to get the updated ciphertext weights and bias.

For multilayer neural networks, the ciphertext backpropagation and ciphertext gradient update are repeated until the first layer is reached [10].

The ciphertext training results in the final ciphertext vector $c' = (u'_{s,1}, u'_{s,2}, \ldots, u'_{s,N})$.

Decryption Cipher Training Results. By decrypting the training results, the final recommendation performance is obtained.

Decrypt($c, q, d, n, N, \mathcal{X}, sk$) → m: Input the ciphertext vector c and the private key vector sk in the input parameters, the known ciphertext vector $c = (u'_{s,1}, u'_{s,2}, ..., u'_{s,N})$, the user's private key vector $sk = (t[1], t[2], \cdots, t[N])$, the inner product of the ciphertext vector c and the user's private key vector sk, and then the modulo q operation is performed on each element of the result, and finally the plaintext vector $m = (u_{s,1}, u_{s,2}, ..., u_{s,N})$ is obtained. It is shown in Fig. 2.

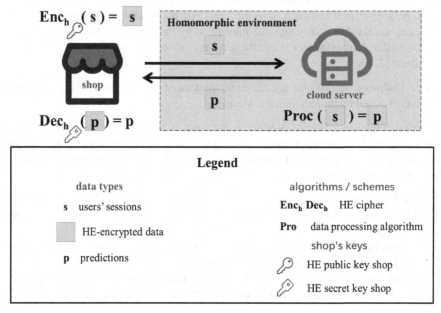

Fig. 2. Session sequences under homomorphic encryption.

3 Position-Aware Graph Neural Network

In this section, we present the position-aware graph neural network (PA-GNN) implemented under BGV homomorphic encryption. The workflow of PA-GNN is illustrated in Fig. 3. The model consists of the following modules: (1) **Graph Model Construction.** In this module, we construct the neighbor-transitional local graph and the position-aware global graph at the local and global levels, respectively. (2) **Item Embedding Learning.** In this module, we utilize the attention mechanism to learn item embeddings at the local level and slices the item embeddings within the current session. Additionally, on the global graph, we separately learn the factor embeddings of different items. (3) **Session Embedding Learning.** In this module, we apply soft-mechanism to the local and global item embeddings to obtain the embeddings of intra-session and inter-session, respectively. These embeddings are then combined to derive the final session representation.

(4) Prediction Layer. In this module, we compare the alternatives with the ground truth items to compute the likelihood of them becoming the next item [11, 12].

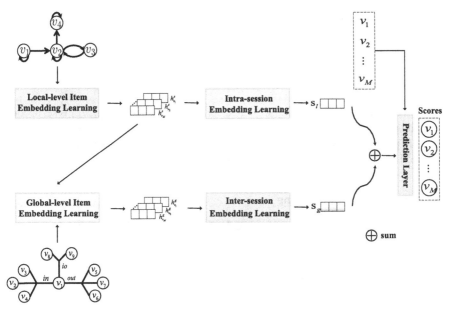

Fig. 3. Workflow of the proposed PA-GNN model.

3.1 Graph Model Construction

We construct a neighbor-transitional local graph $G_s = (\mathcal{V}_s, \mathcal{E}_s)$. Then, we construct four different types of edge which are e_{in}, e_{out}, e_{in-out} and e_{self}. It is shown in Fig. 4.

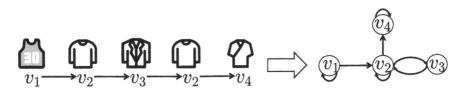

Fig. 4. Neighbor-transitional local graph.

Let $G_{v_i}^g = (\mathcal{V}_g, \mathcal{E}_g)$ be the position-aware global graph. We define $|\mathcal{E}|$ to denote the neighbor-transferring hops. $N_{v_i}^{\mathcal{E}}$ is divided into N_{in}, N_{out} and N_{io}. It is shown in Fig. 5 [13].

A session sequence $s = [v_{s,1}, v_{s,2}, ..., v_{s,\iota}]$ of length ι is known and a position bias $\mathbf{v}_i$ is generated for each element $v_{s,i}$ in the session sequence. Combining these ι positional biases, generates a learned positional matrix $\mathbf{P} = [\mathbf{p}_1, \mathbf{p}_2, ..., \mathbf{p}_\iota]$.

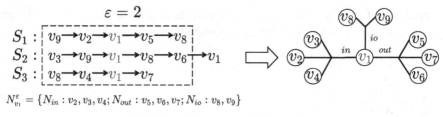

Fig. 5. Position-aware global graph.

For each item $v_{s,i}$ in the session sequence, its positional encoding $\mathbf{p}_{l-i+1}$ is set according to the relative distance between it and the last item $v_{s,l}$. Thus, all the corresponding learnable positional vector $[\mathbf{p}_l, \mathbf{p}_{l-1}, ..., \mathbf{p}_1]$ in the corresponding session sequence are obtained [14].

3.2 Item Embedding Learning

We create a coefficient $\varphi_{i,j}$ to calculate the degree of similarity between neighbor nodes,

$$\varphi_{i,j} = LeakyRelu[\mathbf{W}_{r_{ij}^s}(\mathbf{h}_{v_i} \odot \mathbf{h}_{v_j})], \tag{2}$$

where $r_{i,j}^s \in [in, out, in-out, self]$ indicates the relationship between v_i and v_j.

Next, we apply the *softmax* function for normalization,

$$\varphi_{i,j} = \frac{exp(\varphi_{i,j})}{\sum_{v_x \in N_{v_i}^S} exp(\varphi_{i,x})}. \tag{3}$$

Eventually, we linearly combine them,

$$\mathbf{h}_{v_i}^s = \sum_{v_j \in N_{v_i}^s} \varphi_{i,j}\mathbf{h}_{v_j}. \tag{4}$$

We map the local embedding $\mathbf{h}_{v_i}^s \in \mathbb{R}^d$ in the current session to the K embedding subspaces, obtaining a set $\mathbf{h}_{v_i}^s = [\mathbf{h}_{v_i,1}^s, ..., \mathbf{h}_{v_i,K}^s] \in \mathbb{R}^{\frac{d}{K}}$.

The embedding is mapped into K blocks, each of which represents a factor [15],

$$\mathbf{c}_{i,k} = \frac{\sigma(\mathbf{W}_k^T \cdot \mathbf{v}_i) + \mathbf{b}_k}{||\sigma(\mathbf{W}_k^T \cdot \mathbf{v}_i) + \mathbf{b}_k||_2}. \tag{5}$$

We obtain the average feature of the k th factor of the current session s_k [16],

$$s_k = \frac{1}{|S|} \sum_{v_i \in S} \mathbf{h}_{v_i,k}^S. \tag{6}$$

We combine location information, frequency information and session preference match,

$$\theta_{i,j} = \mathbf{q}_{r_i^g}^{\mathrm{T}} LeakyRelu(\mathbf{W}_{r_{i,j}^g}[\mathbf{s}_k \odot \mathbf{c}_{j,k} \| w_{ij} \| \mathbf{p}_\mu^{r_{i,j}^g}]), \tag{7}$$

$$\mathbf{h}_{N_{r_{i,j}^g}}^k = \sum_{v_j \in N_{r_{i,j}^g}} \theta_{i,j} \mathbf{c}_{j,k}. \tag{8}$$

Finally, we combine these three types of neighbor information.

$$\mathbf{h}_{N_{v_i}^\varepsilon}^k = \mathbf{h}_{N_{in}}^k + \mathbf{h}_{N_{out}}^k + \mathbf{h}_{N_{io}}^k. \tag{9}$$

$$\mathbf{h}_{v_i,k}^g = LeakyRelu(\mathbf{W}_{k_1}[\mathbf{c}_{i,k} \| \mathbf{h}_{N_{v_i}^\varepsilon}]). \tag{10}$$

In the end, we get $\mathbf{h}_{v_i}^g = [\mathbf{h}_{v_i,1}^g, ..., \mathbf{h}_{v_i,K}^g] \in \mathbb{R}^{\frac{d}{K}}$.

3.3 Session Embedding Learning

We embed the corresponding location information,

$$\mathbf{h}_{v_i,k}^{l'} = tanh(\mathbf{W}_{k_2}[\mathbf{h}_{v_i,k}^l \| \mathbf{p}_{t-i+1}^l] + \mathbf{b}_{k_1}), \tag{11}$$

Then, we calculate the average feature of the current session,

$$\mathbf{s}_{f,k}^l = \frac{1}{\iota} \sum_{i=1}^{\iota} \mathbf{h}_{v_i,k}^l. \tag{12}$$

We apply soft-mechanism to calculate the weight coefficient for each item,

$$\gamma_i^l = \mathbf{q}_k^{\mathrm{T}} \sigma(\mathbf{W}_{k_3} \mathbf{h}_{v_i,k}^{l'} + \mathbf{W}_{k_4} \mathbf{s}_{f,k} + \mathbf{b}_{k_2}), \tag{13}$$

$$\mathbf{s}_{l,k} = \sum_{i=1}^{\iota} \gamma_i^l \mathbf{h}_{v_i,k}^l. \tag{14}$$

$\mathbf{s}_l = [\mathbf{s}_{l,1}, \mathbf{s}_{l,2}, \ldots, \mathbf{s}_{l,K}] \in \mathbb{R}^d$ denotes the final intra-session embedding.
We can get the inter-session embedding in a similar way, which is $\mathbf{s}_g = [\mathbf{s}_{g,1}, \mathbf{s}_{g,2}, \ldots, \mathbf{s}_{g,K}] \in \mathbb{R}^{\frac{d}{K}}$.

Linearly combining the intra-session and the inter-session embedding, we obtain the final session representation $\mathbf{S}$,

$$\mathbf{S} = \mathbf{s}_l + \mathbf{s}_g. \tag{15}$$

3.4 Prediction Layer

The recommendation for each item is obtained,

$$\hat{\mathbf{y}}_i = Softmax(\mathbf{S}^T \mathbf{v}_i). \tag{16}$$

Consequently, we use the cross-entropy loss function as the learning objective [17],

$$\mathcal{L}_c = -\sum_{i=1}^{n} \mathbf{y}_i log(\hat{\mathbf{y}}_i) + (1 - \mathbf{y}_i)log(1 - \hat{\mathbf{y}}_i). \tag{17}$$

4 Experiment

4.1 Datasets and Evaluation Metrics

The datasets chosen for this study are Last.fm and Nowplaying. These two datasets are widely used in music recommendation systems. The selected evaluation metrics are P@20 (Precision at 20) and MRR@20 (Mean Reciprocal Rank at 20). These two metrics are commonly used, and higher values indicate better performance of the recommendation model [18].

4.2 Baselines

(1) FPMC: A sequential recommendation model using Markov chains to model user behavior sequences and achieve personalized recommendations by incorporating contextual information and factor decomposition.
(2) GRU4REC: A sequence recommendation model based on GRU (Gated Recurrent Units) used for session data, capturing time-dependent relationships in user behavior sequences for personalized recommendations.
(3) NARM: A neural network-based recommendation model that utilizes attention mechanisms to capture user interest evolution and contextual information, improving personalized recommendation accuracy and effectiveness.
(4) STAMP: A short-term attention/memory priority model in recommendation systems, effectively modeling user short-term interests and long-term preferences by combining attention and memory mechanisms for personalized recommendations.
(5) SR-GNN: A session-based recommendation model based on graph neural networks, representing user behavior sequences as a graph structure and using graph neural networks to learn user interests and contextual information for personalized session-based recommendations.
(6) GCE-GNN: A graph embedding model for learning representations on graph-structured data, utilizing graph convolution and gated contextual excitation to enhance node representations' accuracy and distinctiveness, exhibiting good performance in recommendation systems [19].

4.3 Comparison with Baselines.

To describe the performance of our PA-GNN model, we compared it with other baselines. In each column, we have highlighted the best results for each metric, as shown in Table 2.

It can be observed that our model outperforms the other baselines on both datasets and for both evaluation metrics. This indicates that our model is state-of-the-art.

Table 2. Comparison with baselines.

Method	Last.fm		Nowplaying	
	P@20	MRR@20	P@20	MRR@20
FPMC	16.06	7.32	16.06	7.32
GRU4REC	10.93	5.89	10.93	5.89
NARM	23.3	10.7	23.3	10.7
STAMP	26.47	13.36	26.47	13.36
SR-GNN	25.57	13.72	25.57	13.72
GCE-GNN	33.42	15.42	33.42	15.42
PA-GNN	**35.03**	**16.81**	**35.03**	**16.81**

4.4 Comparison with Factor Number

In order to investigate whether the factor number(K) in PA-GNN can play a role in the performance of our model, we studied the performance of the model with different numbers of factors. We defined the set as $\{1, 2, 4, 8, 16\}$, as shown in Fig. 6.

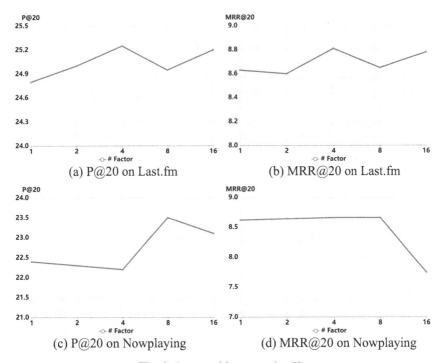

(a) P@20 on Last.fm   (b) MRR@20 on Last.fm

(c) P@20 on Nowplaying   (d) MRR@20 on Nowplaying

Fig. 6. Impact of factor number(K).

Based on our findings, when K = 1, the performance of the model in terms of P@20 and MRR@20 is relatively poor on both datasets. This suggests that a single factor is not sufficient to adequately capture users' preference patterns. As the number of factors increases, the performance gradually improves until it reaches a peak. We observed that the optimal number of factors varies across different datasets: K = 4 on Last.fm and K = 8 on Nowplaying. This indicates that users intents are driven by different factors in different contexts. However, when the number of factors exceeds the threshold, the model's performance tends to decline. This suggests that overly fine-grained preference factors can lead to poorer performance. In conclusion, the selection of an appropriate K plays an essential role in the performance of model.

4.5 Comparisons with Models

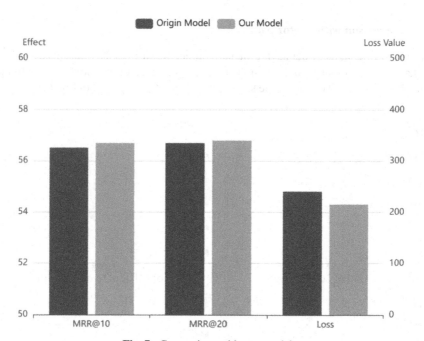

Fig. 7. Comparison with two models.

According to Fig. 7, it is evident that our model outperforms the original model in terms of MRR@10 and MRR@20, and the total loss of this method is smaller than that of the original model. These results indicate that our method delivers better recommendation performance in both small and large datasets compared to the original model. Additionally, the decrease in total loss signifies that under BGV homomorphic encryption, we obtain a more accurate and generalizable graph neural network model, which is beneficial for ensuring the effectiveness and performance of the session-based recommendation system.

5   Conclusion

This paper introduces two models: SSBR and PA-GNN. In SSBR, we employ a session recommendation model that incorporates BGV homomorphic encryption for privacy protection of users. This ensures that user privacy is safeguarded throughout the recommendation process. In PA-GNN, by leveraging graph neural network for item embedding learning and session embedding learning in both local and global contexts, we deliver efficient session recommendations to users. The experimental results using two benchmark datasets demonstrate that, while ensuring user privacy, our proposed model is superior to the state-of-the-art baselines on session-based recommendations (SBRs).

References

1. Gu, P., Han, Y., Gao, W., et al.: Enhancing session-based social recommendation through item graph embedding and contextual friendship modeling. Neurocomputing **419**, 190–202 (2021)
2. Badsha, S., Yi, X., Khalil, I.: A practical privacy-preserving recommender system. Data Sci. Eng. **1**, 161–177 (2016)
3. Soni, K., Panchal, G.: Data security in recommendation system using homomorphic encryption. In: Satapathy, S.C., Joshi, A. (eds.) ICTIS 2017. SIST, vol. 83, pp. 308–313. Springer, Cham (2018). https://doi.org/10.1007/978-3-319-63673-3_37
4. Marcolla, C., Sucasas, V., Manzano, M., et al.: Survey on fully homomorphic encryption, theory, and applications. Proc. IEEE **110**(10), 1572–1609 (2022)
5. Pulido-Gaytan, B., Tchernykh, A., Cortés-Mendoza, J.M., et al.: Privacy-preserving neural networks with homomorphic encryption: challenges and opportunities. Peer-to-Peer Netw. Appl. **14**(3), 1666–1691 (2021)
6. Feng, L., Cai, Y., Wei, E., et al.: Graph neural networks with global noise filtering for session-based recommendation. Neurocomputing **472**, 113–123 (2022)
7. Yagisawa, M.: Fully homomorphic encryption without bootstrapping. Cryptology ePrint Archive (2015)
8. Mono, J., Marcolla, C., Land, G., et al.: Finding and evaluating parameters for BGV. In: El Mrabet, N., De Feo, L., Duquesne, S. (eds.) International Conference on Cryptology in Africa, pp. 370–394. Springer, Cham (2023). https://doi.org/10.1007/978-3-031-37679-5_16
9. Hesamifard, E., Takabi, H., Ghasemi, M.: CryptoDL: deep neural networks over encrypted data. arXiv preprint arXiv:1711.05189 (2017)
10. Yuan, J., Yu, S.: Privacy preserving back-propagation neural network learning made practical with cloud computing. IEEE Trans. Parallel Distrib. Syst. **25**(1), 212–221 (2013)
11. Sang, S., Yuan, W., Li, W., et al.: Position-aware graph neural network for session-based recommendation. Knowl. Based Syst. **262**, 110201 (2023)
12. Li, A., Zhu, J., Li, Z., et al.: Transition information enhanced disentangled graph neural networks for session-based recommendation. Expert Syst. Appl. **210**, 118336 (2022)
13. Chen, Y., Tang, Y., Yuan, Y.: Attention-enhanced graph neural networks with global context for session-based recommendation. IEEE Access **11**, 26237–26246 (2023)
14. Wu, S., Tang, Y., Zhu, Y., et al.: Session-based recommendation with graph neural networks. In: Proceedings of the AAAI Conference on Artificial Intelligence. vol. 33, no. 01, pp. 346–353 (2019)
15. Dong, L., Zhu, G., Wang, Y., et al.: A graph positional attention network for session-based recommendation. IEEE Access **11**, 7564–7573 (2023)

16. Ma, J., Zhou, C., Cui, P., et al.: Learning disentangled representations for recommendation. In: Advances in neural information processing systems, vol. 32 (2019)
17. Li, A., Cheng, Z., Liu, F., et al.: Disentangled graph neural networks for session-based recommendation. IEEE Trans. Knowl. Data Eng. **35**, 7870–7882 (2022)
18. Gwadabe, T.R., Liu, Y.: Improving graph neural network for session-based recommendation system via non-sequential interactions. Neurocomputing **468**, 111–122 (2022)
19. Hidasi, B., Karatzoglou, A., Baltrunas, L., et al.: Session-based recommendations with recurrent neural networks. arXiv preprint arXiv:1511.06939 (2015)

Design of a Fast Recognition Method for College Students' Classroom Expression Images Based on Deep Learning

Jinyun Jiang[1(✉)], Jia Yu[1], and Luying Min[2]

[1] East china jiaotong university, Nanchang, China
tjjjy86@163.com
[2] China Railway International Group Co. Ltd., Nanchang, China

Abstract. The expressions of students in class will directly reflect their classroom state. Therefore, this paper proposes a fast recognition method for college students' classroom expression images based on deep learning. Firstly, the college students' classroom expression images are processed by histogram equalization, graying and clipping to avoid other factors affecting the recognition results of expression images. Secondly, the convolution neural network in deep learning method is used to extract the features of college students' classroom expression images. Finally, support vector machine is used to complete the rapid recognition of college students' classroom expression images. The experimental results show that, compared with the traditional facial expression image recognition methods, this method can improve the recognition accuracy of facial expression images, and shorten the recognition time with an average of 15.1 s.

Keywords: Deep learning · Classroom expression · Expression image · Quick identification

1 Introduction

Classroom expression can reflect students' memory, thinking, perception and other cognitive activities, and play an important role in students' learning motivation and behavior regulation. During the classroom learning process of college students, the expression state will have an impact on the overall learning quality, so solving this problem can effectively improve college students' classroom learning quality [1,2]. However, in the classroom teaching process, a teacher needs to be responsible for managing dozens of students, which makes it difficult for teachers to pay attention to the expression of each student in the classroom [3]. Through modern information technology, the facial expression recognition system can help teachers identify students' facial expressions and give some feedback, so as to strengthen the management of students' classroom learning state, which can

© The Author(s), under exclusive license to Springer Nature Singapore Pte Ltd. 2024
J. Vaidya et al. (Eds.): AIS&P 2023, LNCS 14510, pp. 147–155, 2024.
https://doi.org/10.1007/978-981-99-9788-6_12

improve their learning performance [4]. Therefore, it is necessary to study an effective method of college students' classroom expression image recognition.

Reference [5] proposed a facial expression recognition algorithm based on the attention model. However, this method only denoises the non expression areas, and the accuracy of facial expression recognition is insufficient. Reference [6] proposed a student facial expression recognition method based on chi square distance metric learning. However, this method requires a lot of iterative calculation, which leads to a long time consuming and low efficiency of facial expression recognition. Reference [7] proposed a method of student facial expression recognition based on multi feature combination. However, this method has the problem of insufficient accuracy of facial expression recognition.

In order to solve the problems in the above methods, such as low recognition accuracy and long recognition time, this research proposes a fast recognition method for college students' classroom expression images based on deep learning.

2   Methods

2.1   Preprocessing

Factors such as light and posture will interfere with the results of expression recognition to some extent. The preprocessing of expression images can separate college students' expressions from the background environment of the image, laying the foundation for subsequent feature extraction and rapid expression recognition [8–10]. Histogram equalization can effectively remove the interference caused by light and noise in images, and it's core idea is to broaden the gray level of the image with more pixels and compress the image with less pixels, so as to expand the dynamic range of pixel values, improve the contrast and gray level of the expression image, and improve the clarity of the expression image [11].

For college students' classroom face images, the expression of histogram equalization is:

$$S_k = \sum_{q=0}^{k} \frac{n_q}{N}; k = 0, 1, \ldots, L - 1 \tag{1}$$

In the formula, n_q is the number of pixels whose gray value is q in the expression image, N is the number of pixels in the expression image, and L is the total number of possible gray levels in the image.

Based on the gray level expression image $X(m, n)$ after histogram equalization, feature location of eye muscle group domain of college students' classroom expression image was carried out. Hough transform technology is used to detect the pupil position in college students' classroom expression images [12]. The general expression of Hough transform is:

$$f(X, g) = 0 \tag{2}$$

In the formula, g is the space vector point of Hough transform parameter.

When the points in the image space are transformed into the parameter space, the aggregation of points in the image space can be divided. However, because the advantages of human eyes are not at the same horizontal line, it is necessary to rotate the expression image to correct the angle of the image:

$$[x, y, 1] = [u, v, 1] \begin{bmatrix} cos\theta & sin\theta & 0 \\ -sin\theta & cos\theta & 0 \\ 0 & 0 & 1 \end{bmatrix} \tag{3}$$

In the formula, θ represents the angle of rotation, (x, y) represents the pixel position before rotation, and (u, v) represents the pixel position after rotation [13].

After the rotation processing of college students' classroom expression image is completed, the pupil line of the human eye in the image remains parallel to the horizontal line, then the expression image at this time is considered as a standard image. But the hair, eyelashes and background in the expression image will interfere with expression recognition. In order to avoid these interference factors affecting the final expression recognition, we need to cut the expression image. The schematic diagram of facial expression image inspection is shown in "Fig. 1".

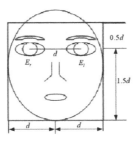

Fig. 1. Expression image clipping model

In "Fig. 1", E_l and E_r are the positions of the left and right pupils respectively. The $2d \times 2d$ area basically contains all types of expressions, so the interference of hair, eyelashes and background in other areas is excluded.

2.2 Feature Extraction

After a series of preprocessing of college students' classroom expression images, the convolutional neural network in deep learning is used to recognize college students' classroom expression images. Convolutional neural network is a feedforward neural network [14], its predecessor is the neural perceptron in the biological vision system. The structure of the neural network is shown in "Fig. 2".

Input the college students' classroom expression images after the above preprocessing into the convolutional neural network for iterative training, so as to

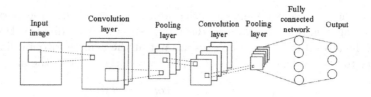

Fig. 2. Basic structure of convolutional neural network

improve the characteristics of the expression images. It can be seen from "Fig. 2" that the convolutional neural network is mainly composed of convolution layer, pooling layer and full connection layer. Different hierarchical structures have different functions. The convolutional layers are used to extract image details and abstract information, achieving parameter sharing; The pooling layer can reduce the size of feature maps and achieve dimensionality reduction; The full connection layer can deeply integrate college students' classroom expression images to output one-dimensional expression image features.

2.3   Fast Recognition

After feature extraction, fast recognition of college students' classroom expression is completed by support vector machine. The classification essence of support vector machine is to divide space r^I into several parts, each part contains a class of samples.

The classification function $g(x)$ of the data set in r^I space is a linear function, and the definition of the classification function is $g(x) = ax + m$. The schematic diagram of Class II linear optimal classification is shown in "Fig. 3".

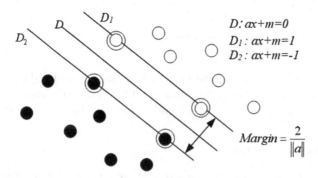

Fig. 3. Schematic Diagram of Class II Linear Optimal Classification

From "Fig. 3", it can be concluded that the classification interval can be expressed as $2/\|a\|$.

If the two-dimensional classification space is extended to the I-dimensional space, the linear separable problem of r^I can be transformed into a quadratic programming problem of a and b, as follows:

$$\min_{a,m} \frac{1}{2}||a||^2 \tag{4}$$

By introducing a series of operations such as Lagrangian function and relaxation variables in formula "(4)", we can get:

$$L(a,m,\varepsilon,\partial,\beta) = \frac{1}{2}||a||^2 + C\sum_{i=1}^{i}\varepsilon_i - \sum_{i=1}^{i}\beta_i\varepsilon_i$$
$$- \sum_{i=1}^{i}\partial_i(y_i((ax_i)+m)-1+\varepsilon_i) \tag{5}$$

At this time, the dual problem can be expressed as:

$$\max_{\partial,\beta} -\frac{1}{2}\sum_{i=1}^{i}\sum_{j=1}^{j}y_iy_j\partial_i\partial_j(x_ix_j) + \sum_{i=1}^{i}\partial_j \tag{6}$$

In the above formula, ∂ represents linear decomposability, β represents the Lagrange multiplier, ε is the penalty coefficient and C is the value of relaxation variable. Then there must be solutions ∂ and β.

Through the above calculation, the output of the solution is the result of college students' facial expression image recognition. Support vector machine is used to recognize college students' classroom expressions.

3   Experimental Verification

3.1   Experimental Data

In order to collect the real classroom expressions of college students, 22 students from a university were selected as the research objects in this study, including 15 girls and 9 boys. With the permission of the 22 students, their spontaneous expression data in the classroom learning process were recorded.

The recording equipment for college students' classroom expressions is the Inter (R) Real Sense (TM) Camera SR300 camera. Two learning videos of different difficulty are used for students to learn in the emotional induction stage. Each course is equipped with 45 min of test time. The students were recorded to watch the learning video and the expression changes during the test. Finally, 900 peak expression images were obtained from 22 test students. The main expression categories were surprise, fatigue, confusion, happiness and center. College students' classroom expression image samples (part) are shown in "Fig. 4".

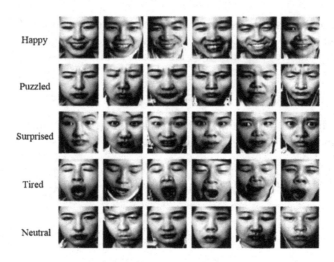

Happy

Puzzled

Surprised

Tired

Neutral

Fig. 4. Image samples (part)

3.2 Test Plan

The above process has completed the collection of sample data. The overall experimental scheme is set as follows: take the accuracy of expression image recognition and the time consumption of expression image recognition as indicators, and compare and verify the method in this paper with the method in Reference [5] and the method in Reference [6].

3.3 Analysis of Experimental Results

Expression Image Recognition Accuracy. The higher the recognition accuracy of expression image, the stronger the recognition performance of the method, and the accurate expression state of college students' classroom learning can be obtained. Taking the recognition accuracy of facial expression image as an indicator, the recognition accuracy results of this method are compared with those of two traditional literature methods. The comparison results of expression image recognition accuracy of the three methods are shown in "Fig. 5".

It can be seen from the results shown in "Fig. 5" that, compared with the two traditional literature methods, the recognition accuracy of facial expression images in this method is significantly improved, and the recognition accuracy curve is relatively stable, which always remains at the level of about 95%, with the highest recognition accuracy reaching 97%. On the other hand, although the identification accuracy curves of the two literature methods show an upward trend of fluctuation, the overall level is low, with the highest identification accuracy of about 80%, and there is still much room for improvement.

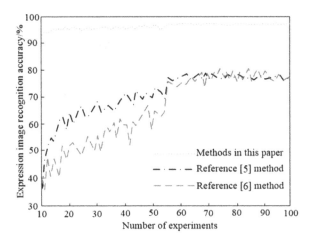

Fig. 5. Recognition accuracy of expression image

Time Consuming for Expression Image Recognition. Due to the large number of college students, the number of expression images generated is relatively rich. In order to complete the recognition of college students' classroom expression images as soon as possible, higher requirements are put forward for the recognition efficiency of different methods of expression images. In order to meet the requirements of large-scale expression image recognition, the recognition time consumption of different methods is tested to fully verify the performance of this method. Table 1 shows the comparison results of expression image recognition time consumption of the three methods.

It can be seen from Table 1 that under the multiple comparative experiments of the three methods, the time consumption of facial expression image recognition of this method is significantly reduced, and the longest recognition time of this method is not more than 17 s. The average recognition time of this method under multiple experiments is 15.1 s. The identification time of reference [5] method is always more than 40 s, and the average identification time is 46.3 s; However, the single recognition time consumption of reference [6] method is the highest among the three methods, reaching 58 s, and the average recognition time consumption of this method is 46.5 s. Therefore, this method can shorten the recognition time of about 30 s, further improve the recognition efficiency of classroom expression images, and better meet the requirements of college students' classroom expression image recognition.

Table 1. Time consumption

Number of experiments	Time consuming for identification/s		
	Methods in this paper	Reference [5] method	Reference [6] method
10	15	45	30
20	14	47	22
30	16	43	54
40	14	44	37
50	13	49	58
60	17	46	53
70	14	48	52
80	17	47	54
90	16	49	56
100	15	45	49
Average value	15.1	46.3	46.5

4 Conclusion

In order to improve the quality of college students' classroom learning and effectively supervise their classroom learning status, this paper proposes a method of college students' classroom expression image recognition based on deep learning. The performance of the method is verified from both theoretical and experimental aspects. In the process of college students' classroom expression image recognition, the method has higher expression image recognition accuracy and shorter expression image recognition time. Specifically, compared with the recognition method based on attention model, the recognition accuracy of the expression image in this method is significantly improved, with the highest recognition accuracy reaching 97%; Compared with the recognition method based on chi square distance measurement learning, the recognition time of this method is significantly shortened, and the average recognition time is 15.1 s. Therefore, this method can better meet the requirements of college students' classroom expression image recognition.

References

1. LI, S., Gao, Y., Wang, F., SHI, T.: Research on classroom expression recognition based on deep circular convolution self-encoding network. In: 2020 15th International Conference on Computer Science & Education (ICCSE), pp. 523–528. IEEE (2020)

2. Pabba, C., Kumar, P.: An intelligent system for monitoring students' engagement in large classroom teaching through facial expression recognition. Expert. Syst. **39**(1), 12839–12852 (2022)

3. Yu, W., Liang, M., Wang, X., Chen, Z., Cao, X.: Student expression recognition and intelligent teaching evaluation in classroom teaching videos based on deep attention network. J. Comput. Appl. **42**(3), 743 (2022)

4. Pan, X., Jian, C., Ma, R.: Classroom teaching feedback system based on facial expression recognition. Comput. Syst. Appl. **30**(10), 102–108 (2021)

5. Chu, J., Tang, W., Zhang, S., Lv, W.: An attention model-based facial expression recognition algorithm. Laser Optoelectron. Progress **57**(12), 197–204 (2020)

6. Qin, Y., Zhao, E.: Facial expression recognition algorithm based on chi-squared distance metric learning. Comput. Eng. Design **43**(5), 1412–1418 (2022)

7. Li, Z., Li, K., Deng, J.: Research on facial expression recognition method based on multi-feature combination. Appl. Res. Comput. **S02**, 398–400 (2020)

8. Quan, Y.: Development of computer aided classroom teaching system based on machine learning prediction and artificial intelligence KNN algorithm. J. Intell. Fuzzy Syst. **39**(2), 1879–1890 (2020)

9. Bhatti, Y.K., Jamil, A., Nida, N., Yousaf, M.H., Viriri, S., Velastin, S.A.: Facial expression recognition of instructor using deep features and extreme learning machine. Comput. Intell. Neurosci. **2021**, 1–17 (2021)

10. Ashwin, T., Guddeti, R.M.R.: Affective database for e-learning and classroom environments using Indian students faces, hand gestures and body postures. Futur. Gener. Comput. Syst. **108**, 334–348 (2020)

11. Zheng, K., Yang, D., Liu, J., Cui, J.: Recognition of teachers facial expression intensity based on convolutional neural network and attention mechanism. IEEE Access **8**, 226437–226444 (2020)

12. Huang, Q., Liu, J.: Practical limitations of lane detection algorithm based on Hough transform in challenging scenarios. Int. J. Adv. Rob. Syst. **18**(2), 1–13 (2021)

13. Fu, Y.: Facial pose estimation method on single image based on facial feature points. Comput. Eng. **47**(4), 197–203+210 (2021)

14. Sun, Y., Xu, J., Wu, H., Lin, G., Mumtaz, S.: Deep learning based semi-supervised control for vertical security of maglev vehicle with guaranteed bounded airgap. IEEE Trans. Intell. Transp. Syst. **22**(7), 4431–4442 (2021)

Research on ALSTM-SVR Based Traffic Flow Prediction Adaptive Beacon Message Joint Control

Yuqing Zeng[1]([✉]) [ID], Botao Tu[2] [ID], Guoqing Zhong[1] [ID], and Dan Zou[1] [ID]

[1] School of Information Engineering, East China Jiaotong University,
Nanchang 330013, China
zyq20010116@163.com, 1405@ecjtu.edu.cn

[2] Jiujiang Management Centre Information Sub-centre, Jiangxi Transportation
Investment Group, Jiujiang 332020, China

Abstract. In Vehicular Ad Hoc Networks (VANETs), vehicles rely on periodic broadcasting of beacon messages to perceive their surrounding environment. The interaction between vehicles' Basic Safety Messages (BSMs) allows drivers to promptly gather information about the surrounding traffic conditions, enabling them to identify potential hazards and prevent accidents. However, effectively distributing beacon messages in complex and dynamic traffic environments presents a significant challenge. To address this challenge, this paper proposes an adaptive joint control scheme for beacon message dissemination based on an improved attention mechanism, specifically, the combination of Long Short-Term Memory (LSTM) and Support Vector Regression (SVR), for traffic flow prediction. The scheme leverages historical traffic flow data as input and applies various data preprocessing techniques, including reconstruction and normalization. By employing an LSTM model with an improved attention mechanism and denormalization, accurate predictions of traffic flow can be achieved. The predicted parameters are then utilized for the joint control of beacon message transmission. Simulation results demonstrate that the proposed joint control scheme significantly enhances network performance by effectively reducing channel congestion. The approach provides improved reliability and efficiency in distributing beacon messages, contributing to the overall safety and effectiveness of VANETs.

Keywords: Internet of vehicles · LSTM-SVR model · traffic flow prediction · adaptive transmission · joint control

1 Introduction

Vehicular Ad-hoc Networks (VANETs) are an important component of intelligent transportation systems, primarily relying on communication to achieve

Supported by organizations of the National Natural Science Foundation of China, Jiangxi Provincial Natural Science Foundation and Science and Technology Research Project of Jiangxi Education Department.

information sharing between vehicles, which has gradually become an important means of solving various traffic problems [1]. Vehicle-to-vehicle (V2V) communication can effectively improve the safety of vehicle driving, and prevent and control events such as vehicle congestion and accidents. In VANETs, the basic information unit is the vehicle equipped with an On Board Unit (OBU), which can collect its own information and exchange information with Road Side Units (RSU) or neighboring vehicles through wireless networks. Figure 1 shows the four communication methods in the connected vehicle network, which are V2V, V2R, R2R, and V2V-V2R [2,3]. V2V communication achieves the most basic information exchange between vehicles and relies on the installed OBUs. The OBU is an integrated device that includes a vehicle positioning system and a vehicle communication system [4,5]. Road side unit communication (V2R) involves both the vehicle-mounted unit and the roadside unit deployed infrastructure. The roadside unit can receive vehicle status information, distribute it, or pass it to a server for further processing. The roadside units communicate with other infrastructure through cloud servers and RSU-to-RSU (R2R) communication.

In the 802.11p standard, the vehicle periodically broadcasts beacon messages through the control channel (CCH). However, due to the high dynamics of vehicular Ad-hoc Networks, the channel load fluctuates greatly, lead to channel network congestion, which makes the message unable to be distributed in time, and even threatens the safety of drivers. Therefore, researchers adjust the corresponding transmission parameters to offset the congestion [6]. Most of the existing beacon message transmission mechanisms are optimized by adjusting the current parameters. In fact, this optimization method has a certain lag for the vehicular Ad-hoc Networks whose topology changes rapidly. Especially when the interval between messages is large, it will be more obvious. This will lead to a decrease in channel utilization or channel congestion.

The research of beacon transmission mechanism generally focuses on two aspects: the transmission rate of beacon and the transmission power of beacon [7,8]. The rate of periodic security message generation has a significant impact on security applications. With a high message generation rate, the update rate of vehicles receiving messages will increase, and the vehicle's perception accuracy to the surrounding environment will also improve, but this will bring the problem of channel congestion [9]. However, if the message generation rate is too low, the timeliness of the message will be reduced and sufficient accuracywill not be provided for secure applications.

Therefore, it is necessary to control the generation rate of periodic secure messages adaptively and adopt the generation rate that is most suitable for the current communication environment. If each vehicle node adopts fixed transmission power in the vehicular Ad-hoc Networks, the transmission range of its beacon message is also fixed, which is obviously not suitable for the vehicle node topology that changes at any time [10]. At the same time, a higher transmission power means a larger transmission distance, and the signal transmission has a higher robustness. It will also cause interference to other nearby wireless signal transmission. On the contrary, the lower transmission power has a smaller transmission distance, and it is more susceptible to the interference of the surround-ing environment, and the signal robustness is poor. Therefore, we can find that a single optimized beacon message generation frequency or beacon message transmission power can not meet the requirements of all scenarios of beacon message sending. So this paper designs an adaptive beacon transmission rate and power joint control scheme based on traffic flow prediction. The ALSTM-SVR model is used to forecast the traffic flow based on the historical traffic flow time series data. After the predicted parameters are obtained, Combined with the joint optimization control of beacon message generation frequency and transmission power, the optimization of beacon message transmission performance is realized.

2   Experiments

2.1   Traffic Flow Prediction and Joint Control Schemes

As an important part of the intelligent transportation system, VANET is a special mobile ad hoc network, VANET is characterized by: changeable network topology, high-speed mobility of nodes, link instability, strong node storage capacity, and limitations of node transmission range. Considering that beacon messages contain highly predictable vehicle movement information, and the vehicle is surrounded by dynamic environmental noise, beacon messages are inherently uncertain. When traffic density increases, the number of collisions on the channel also increases, negatively affecting message accuracy.

Therefore, the scheme is divided into two parts, one is to predict traffic flow based on historical traffic data, and the prediction results can be stored in all vehicles. After that, the vehicle is running on the road, the channel load is monitored, the threshold is set by the predicted parameters of the future moment, and when the threshold is exceeded, the adaptive transmission scheme is initiated, and the previously predicted parameters are used to adaptively control the beacon.

2.2   LSTM-SVR Combined Model for Traffic Flow Prediction

The LSTM algorithm has strong temporal characteristics and can fully mine the rules of historical data, but requires a large number of samples. In LSTM, the self-attention mechanism is used to calculate the similarity between the current time step and the previous time steps at each time step, resulting in a context vector. This vector weights historical information and can be added to the input at the current time step to generate a new input vector. This makes the model more flexible in using historical information to generate output.

The SVR algorithm has the advantages of nonlinear mapping and small sample learning, but has poor temporal characteristics. However, the samples we need to predict have both temporal and small sample properties, so we consider combining the two.

In this paper, preprocessed traffic flow data is input into an LSTM model with improved attention mechanism for training, which consists of an input layer, two LSTM layers, a self-attention layer, a dropout layer, and a dense layer with a sigmoid activation function. Similarly, preprocessed traffic flow data is input into an SVR model for training. After obtaining the trained LSTM and SVR prediction models, we input data from multiple time points into the ALSTM prediction model to obtain predicted traffic flow data. We then take the average of the data and the traffic flow data from the previous time step and input it into the SVR prediction model for training, and finally obtain the results. The specific neural network model is depicted in Fig. 1.

Thus, we can obtain the vehicle density di+1 of the predicted future traffic state.

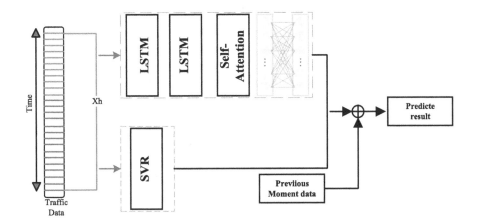

Fig. 1. Neural network structure diagram

To assess the effectiveness of the models, a comparative analysis is conducted, commonly used evaluation metrics are employed in the prediction models: EVS (Explained Variance Score), Mean Absolute Error (MAE), Mean Squared Error (MSE), Root Mean Squared Error (RMSE), R2. The computational formulas are illustrated as Eqs.(1)(2)(3)(4)(5).

$$EVS = 1 - \frac{Var(y - \hat{y})}{Var(y)} \tag{1}$$

$$MAE = \frac{1}{n} \sum_{i=1}^{n} |(y_i - \hat{y}_i)| \tag{2}$$

$$MSE = \frac{1}{n} \sum_{i=1}^{n} (y_i - \hat{y}_i)^2 \tag{3}$$

$$RMSE = \sqrt{\frac{1}{n} \sum_{i=1}^{n} (y_i - \hat{y}_i)^2} \tag{4}$$

$$R^2 = 1 - \frac{\sum_{i=1}^{n} (y_i - \hat{y}_i)^2}{\sum_{i=1}^{n} (y_i - \bar{y})^2} \tag{5}$$

The experimental results, as shown in Fig. 2, demonstrate the performance of the proposed improved LSTM-SVR model. Compared to the baseline models LSTM and GRU, the ALSTM-SVR model exhibits varying degrees of improvement in parameters such as EVS, Mae, Mse, Rmse, and R2. The higher accuracy achieved by the ALSTM-SVR model in predicting traffic flow data correlates with its enhanced performance. The specific experimental results are summarized in Table 1.

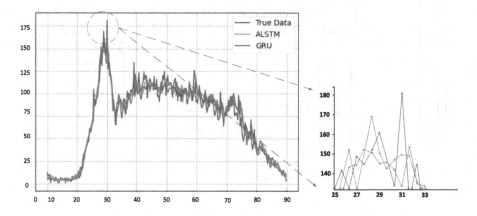

Fig. 2. Predicted Results Figure

2.3   Joint Control Scheme

The higher the beacon message generation frequency is, the more beacon messages will be received by neighboring vehicles, and the higher the vehicle perception accuracy of traffic environment will be, but the higher the beacon message generation frequency may bring about the problem of channel congestion. However, if the generation frequency of beacon message is low, the real-time performance of the message will be reduced, which may not meet the real-time requirements of some security applications. As for transmission power, usually the security application only needs to meet the requirement that neighbor vehicles within the range of one hop can transmit beacon messages to each other. In the freeway environment with low vehicle density, the transmission power

Table 1. Comparison of experimental structure data

	LSTM	GRU	ALSTM-SVR
EVS	0.93	0.93	0.94
MAE	7.34	7.20	7.21
MSE	98.88	99.32	95.8
RMSE	9.94	9.94	9.97
R2	0.93	0.93	0.94

of beacon messages can be increased to achieve coverage adjustment of beacon messages without causing channel congestion. In thecongested road with high vehicle density, the lower transmission power of beacon message can ensure the transmission of beacon message between adjacent vehicles, and effectively prevent channel congestion. Therefore, we consider the combination of the two for joint control, and the specific scheme is as follows.

The scheme uses the vehicle-mounted AD hoc Network mobility Simulator Sumo, Network Simulator 2 (NS2) and Python co-simulation to analyse the performance of the proposed adaptive transmission scheme. Sumo is mainly responsible for vehicle mobility simulation and provides mobility files for simulator NS2. Python combination model is mainly used to predict shortterm traffic flow parameters. NS2 mainly realizes the simulation of vehicular Ad-hoc Networks communication.

3 Results

In order to analyze the performance of the proposed joint adaptive control method, this paper divides it into two groups of experiments and compares them with those without adaptive control. The network performance parameters evaluated are as follows: (1) distribution delay, which represents the average time delay of beacon message accessing channel. This metric is particularly important for safety-critical applications where communication delays between vehicles in the Internet of vehicles must be kept to a very low range. (2) Packet Collision Rate. (3) Channel Busy Ratio (CBR), which is defined as the proportion of the time when channel detection is in busy state within a certain monitoring time T. The experimental results are shown below.

Figure 3 (a) illustrates the variation of distribution delay with density. Under fixed transmission power and period, the distribution delay rapidly increases with the increasing vehicle density due to the increased number of nodes contending for the channel, resulting in severe network congestion. However, when using our proposed algorithm, we dynamically control the transmission rate and power to adapt to congestion situations. As a result, the delay growth becomes smoother. Through comparison, we observed a significant improvement in packet delivery rate and a notable decrease in distribution delay compared to the fixed power and period method as well as the power control method. Figure 3 (b) exhibits a similar trend, with a more pronounced effect observed when the vehicle density exceeds 40 vehicles/km. This demonstrates that our adaptive solution successfully enhances the performance of beacon message transmission and effectively reduces channel congestion.

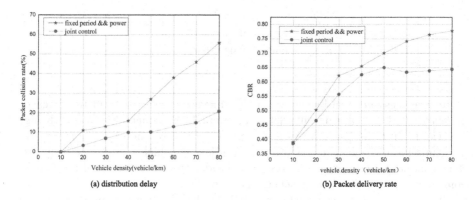

(a) distribution delay   (b) Packet delivery rate

Fig. 3. Shows the variation of distribution delay with density.

4   Conclusion

This paper proposes a joint algorithm for beacon power and period control based on vehicle traffic flow prediction. By enhancing the attention mechanism in the LSTM-SVR composite model, we analyze and predict historical traffic flow data with high accuracy, obtaining reliable predictive parameters. We dynamically control the beacon transmission power and period, taking into account the limited channel resources. We assign appropriate weights to the transmission power and period, enabling adaptive adjustment of beacon period and power in different traffic conditions. Through simulation analysis, our algorithm significantly improves the stability of the communication link. Experimental results demonstrate the effectiveness of our method in enhancing network performance. In future research, we plan to consider the topology in high-speed scenarios and investigate the impact of speed and acceleration on the adaptive beacon scheme, taking into account factors beyond vehicle density.

Acknowledgements. This work was supported in part by the National Natural Science Foundation of China under Grant 92159102, Grant 11862006, Grant 61862025, in part by the Jiangxi Provincial Natural Science Foundation under Grant 20202BAB205011, in part by Science and Technology Research Project of Jiangxi Education Department(GJJ209928).

References

1. Devi, M.S., Malar, K.: Improved performance modeling of intelligent safety message broadcast in vanet: a survey. In: 2014 International Conference on Intelligent Computing Applications, pp. 95–98. IEEE (2014)

2. Kaiwartya, O., et al.: Internet of vehicles: motivation, layered architecture, network model, challenges, and future aspects. IEEE Access **4**, 5356–5373 (2016)
3. Baldessari, R., et al.: Car-2-car communication consortium-manifesto. (2007)
4. Liang, L., Peng, H., Li, G.Y., Shen, X.: Vehicular communications: a physical layer perspective. IEEE Trans. Veh. Technol. **66**(12), 10647–10659 (2017)
5. Peng, H., Liang, L., Shen, X., Li, G.Y.: Vehicular communications: a network layer perspective. IEEE Trans. Veh. Technol. **68**(2), 1064–1078 (2018)
6. Sospeter, J., Di, W., Hussain, S., Tesfa, T.: An effective and efficient adaptive probability data dissemination protocol in VANET. Data **4**(1), 1 (2018)
7. Wang, M., Chen, T., Du, F., Wang, J., Yin, G., Zhang, Y.: Research on adaptive beacon message transmission power in VANETs. J. Ambient Intell. Humanized Comput. **13**, 1307–1319 (2022)
8. Chang, H., Song, Y.E., Kim, H., Jung, H.: Distributed transmission power control for communication congestion control and awareness enhancement in VANETs. PloS one **13**(9), e0203261 (2018)
9. Hassan, A., Ahmed, M.H., Rahman, M.A.: Adaptive beaconing system based on fuzzy logic approach for vehicular network. In: 2013 IEEE 24th Annual International Symposium on Personal, Indoor, and Mobile Radio Communications (PIMRC), pp. 2581–2585. IEEE (2013)
10. Xie, Y., Ho, I.W.-H., Magsino, E.R.: The modeling and cross-layer optimization of 802.11 p VANET unicast. IEEE Access **6**, 171–186 (2017)

An Improved Hybrid Sampling Model for Network Intrusion Detection Based on Data Imbalance

Zhongyuan Gong⬚, Jinyun Jiang⬚, Nan Jiang⬚, and Yuejin Zhang(✉)⬚

East China Jiaotong University, Nanchang, China
jiangnan1018@acm.org, zyjecjtu@foxmail.com

Abstract. Network intrusion detection constitutes a pivotal element in safeguarding computer networks against malicious attacks and unauthorized access. With the widespread proliferation of the internet and the rapid evolution of information technology, network intrusions have become increasingly commonplace and intricate, underscoring the growing significance of network intrusion detection. In order to enhance the performance of network intrusion detection, we propose a sampling method that combines ADASYN with GMM, denoted as AGM. Firstly, the dataset undergoes preprocessing to mitigate noise, inconsistencies, and incompleteness issues inherent in the UNSW-NB15 dataset. Subsequently, sampling processing is applied to the dataset to mitigate the bias towards predicting majority classes, thereby improving prediction accuracy for minority classes. In conclusion, the amalgamation of CNN, BiLSTM, and Channel-Attention in the refined network architecture, CNN-BiLSTM-ATT, capitalizes on the distinctive advantages of each component. The classification outcomes of our experiments demonstrate the notable efficacy of the enhanced sampling technique and network structure for the task of network intrusion detection.

Keywords: Network intrusion detection · Deep learning · Class imbalance · UNSW-NB15

1 Introduction

As the internet becomes more pervasive and information technology advances rapidly, network intrusions have grown in both frequency and complexity. Malicious attacks can lead to interruptions in network services, affecting business continuity and stability. Furthermore, they may result in substantial economic losses due to data loss, leakage, and destruction. As a result, network intrusion detection has assumed increasing significance. In recent years, machine learning techniques [1–3] have gained widespread traction in the realm of network intrusion detection. By training models to discern patterns between normal network

---

This paper was supported by the National Natural Science Foundation of China.

traffic and malicious behaviors, machine learning has demonstrated its capacity to adapt effectively to emerging attack vectors. Furthermore, with the ascent of deep learning, neural networks have made substantial advancements in the field of network intrusion detection. Deep learning methodologies can autonomously extract features from raw data and acquire intricate patterns from vast datasets. Recent investigations [4] have underscored the superior performance of such deep learning-based Network Intrusion Detection Systems (NIDS) when confronted with substantial datasets.

In this study, our primary objective is to address the issue of dataset imbalance. In scenarios characterized by imbalanced data, models often exhibit a bias towards predicting the majority class, resulting in suboptimal performance for minority classes. A balanced dataset helps mitigate this bias, ensuring equitable treatment of all categories by the model. To tackle this data imbalance challenge, we have refined a sampling approach that combines ADASYN with GMM. Our research indicates that this sampling method leads to a discernible improvement in both model training and final classification outcomes. Moreover, we have devised a network data preprocessing technique tailored to the deep hierarchical network model proposed herein, tailored to combat complex, multidimensional network threats. This preprocessing method effectively mitigates noise, inconsistencies, and data incompleteness inherent in the UNSW-NB15 dataset. Finally, we have enhanced a network architecture that integrates CNN, BiLSTM, and Channel Attention. Experimental results unequivocally demonstrate the substantial performance enhancements in network intrusion detection conferred by this network structure.

2   Related Research

In 2017, Chuanlong Yin [4] proposed a Recurrent Neural Network Intrusion Detection System (RNN-IDS). Research has shown that RNN-IDS is well-suited for building high-precision classification models and outperforms traditional machine learning classification methods in both binary and multiclass classification scenarios. In 2019, Yong Zhang [5] and colleagues introduced a novel deep learning intrusion detection network known as Parallel Cross Convolutional Neural Network (PCCN). PCCN integrates traffic features learned from two parallel Convolutional Neural Networks (CNNs) and proves to be effective in leveraging a small number of samples, thereby enhancing the detection performance for imbalanced anomalous traffic. Additionally, Kaiyuan Jiang [6] proposed a Deep Hierarchical Network, which is an innovative intrusion detection model that integrates an improved LeNet-5 and LSTM neural network structure. This model can simultaneously learn the spatial and temporal features of network traffic, and it exhibits superior performance compared to other intrusion detection models, achieving optimal detection accuracy. In 2022, Wai Weng Lo [7] improved Graph Neural Networks (GNNs) to capture edge features and topology information for Internet of Things (IoT) network intrusion detection using flow-based data. This research demonstrates the significant potential of GNNs in the field of network intrusion detection, providing motivation for future studies. Furthermore, Yanfang Fu [8] and their team addressed the issue of imbalanced datasets by employing an Adaptive Synthetic Sampling (ADASYN) method for oversampling minority class samples. They also utilized an enhanced Stacked Autoencoder for data dimensionality reduction to enhance information fusion. Experimental results show that these improvements effectively mitigate data imbalance issues, making the dataset relatively more balanced. These studies offer valuable insights into deep learning approaches for intrusion detection and solutions for handling data imbalance. They are expected to provide essential directions for future research in the field of network intrusion detection. In this article, we primarily focus on improving an ADASYN combined with Gaussian Mixture Model (GMM) hybrid sampling method to address data imbalance. Additionally, we integrate a network structure involving CNN, BILSTM, and Channel Attention. Experimental results demonstrate the effectiveness of this hybrid sampling method in mitigating data imbalance issues.

3   Dataset

3.1   UNSW-NB15

The UNSW-NB15 dataset [9,10], developed by the University of New South Wales (UNSW), serves as a comprehensive simulation of various malicious activity types in a real-world network environment, alongside corresponding normal network traffic. This dataset comprises a total of 2.54 million network traffic samples, each characterized by 49 features, and encompasses 9 distinct attack categories. It is worth noting that normal traffic constitutes 87.35% of the dataset, while attack traffic represents 12.65%. Furthermore, within the category of attack traffic, there exists a significant imbalance among different attack types. To create distinct sets for training, testing, and validation, we partitioned the original UNSW-NB15 dataset according to a 7:2:1 ratio.

3.2   Dataset Preprocessing

While the UNSW-NB15 dataset is designed to emulate real-world network conditions, it may still exhibit certain imperfections such as noise, inconsistencies, and incompleteness. Anomalies within the dataset, which can result from measurement errors, data entry inaccuracies, or other factors, should also be acknowledged. Recognizing and rectifying these anomalies is essential to prevent their potential impact on the accuracy of subsequent analyses. Data cleansing procedures play a pivotal role in enhancing the dataset's quality, establishing a more dependable foundation for subsequent research and analyses. One-Hot Encoding, a widely employed data preprocessing technique, is utilized to convert categorical variables into binary vector representations. Its principal advantage when dealing with categorical attributes is its ability to preserve their unordered nature. Through One-Hot Encoding, categorical features are transformed into binary representations, enabling algorithms to directly handle these features. Following the application of One-Hot Encoding, the feature dimensionality of the UNSW-NB15 dataset expanded from 47 to 208.

Feature simplification, as outlined in reference [11], entailed the removal of redundant and irrelevant attributes from the UNSW-NB15 dataset. Specifically, six features - 'srcip,' 'sport,' 'dsport,' 'dstip,' 'ltime,' and 'stime' - were excluded. These six attributes were identified as redundant, potentially offering information closely related to other features. Through feature simplification, the removal of these redundant features resulted in a reduced feature dimensionality, thereby improving algorithmic performance and efficiency while retaining crucial information. Subsequent to feature simplification, the dataset's dimensionality was reduced to 202.

As the final step in data preprocessing, data standardization was applied to normalize values across different features to a consistent scale. This standardization process, vital for numerous machine learning algorithms, mitigates dimensional variations among features, thereby enabling more stable and accurate model training.

3.3   Class Imbalance Processing

The UNSW-NB15 dataset presents an extreme class imbalance, with the 'Worms' category accounting for only 0.05% of the samples, and the 'Shellcode,' 'Backdoors,' and 'Analysis' attack traffic categories each comprising less than 1%. In this research, we have enhanced the Synthetic Minority Over-sampling Technique known as ADASYN, by combining it with a Gaussian Mixture Model (GMM)-based clustering under-sampling method as introduced in SGM [8,12,13]. ADASYN represents an evolution of the SMOTE technique, retaining its key advantages, such as preserving the original sample information and addressing class boundary challenges. It employs SMOTE's interpolation technique to expand the dataset by generating new samples between instances of the minority class and their neighboring majority class counterparts. ADASYN operates as an adaptive oversampling approach, generating synthetic samples based on the dataset's density. By taking into account the density information of minority class instances, ADASYN prioritizes the generation of synthetic samples for minority instances situated in lower-density regions. This targeted synthesis approach enhances the model's ability to recognize minority class instances. Initially, we employ the ADASYN algorithm for oversampling the minority class samples, generating synthetic instances. ADASYN calculates the number of synthetic samples to create for each minority class instance based on the quantity of neighboring samples in its proximity, thus maintaining the sample distribution density.

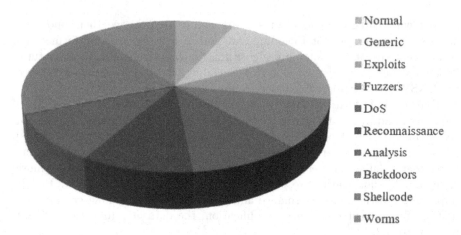

Fig. 1. Dataset after undergoing AGM sampling

Next, the original minority class samples are combined with the synthetically generated samples, resulting in an expanded dataset for the minority class. Subsequently, Gaussian Mixture Modeling (GMM) is applied to this augmented minority class dataset to model and estimate the data distribution. Finally, leveraging the estimated GMM model, new samples can be generated from each Gaussian distribution. These newly generated samples effectively fill gaps within the data space of the original data distribution, thereby further enhancing the model's generalization capability. Figure 1 illustrates the distribution of data categories after the combined sampling process involving ADASYN and GMM.

Figure 2 displays a bar chart illustrating the dataset comparison of UNSW-NB15 after undergoing SGM and AGM sampling. Following AGM sampling, the dataset exhibits a notably larger volume of data, thereby augmenting the representation of minority class samples. This augmentation enriches the model's feature learning capabilities and bolsters its resilience. These benefits facilitate the utilization of more complex network architectures.

4   Network Architecture

The improved CNN-BiLSTM-Attention architecture in this study combines the strengths of CNN and BiLSTM models. CNN excels in detecting local features, while BiLSTM effectively captures temporal relationships within sequential data, making it particularly well-suited for modeling sequential data in network intrusion detection, such as network connection traffic.

The Channel Attention mechanism adaptively adjusts the weights for different channels (feature maps), enhancing the representation of crucial features. This is highly effective for feature selection and extraction in network intrusion detection, allowing the model to focus more on the information in attack traffic. Figure 3 illustrates the specific network structure. Processed data first undergoes

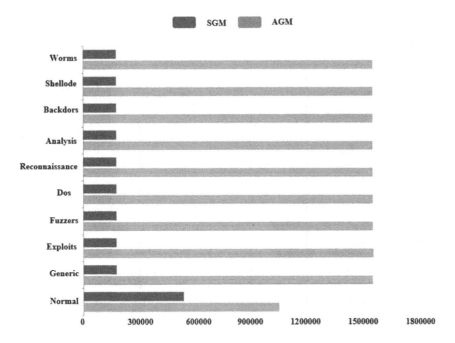

Fig. 2. Dataset comparison after undergoing both SGM and AGM sampling

a 1D convolutional layer with 32 convolutional kernels of size 3, utilizing ReLU activation and padding to extract local features. Subsequently, the second convolutional layer with the same configuration is employed, followed by max-pooling to reduce feature map dimensions and a 0.2 dropout layer to prevent overfitting. Batch Normalization is applied afterward for standardization. Next, the data is processed through a third 1D convolutional layer with 64 convolutional kernels of size 3, again utilizing ReLU activation and padding. This is followed by a fourth 1D convolutional layer with the same configuration, employing 64 convolutional kernels and ReLU activation, along with padding. Max-pooling and a 0.2 dropout layer are applied once more, followed by Batch Normalization. The data then passes through multiple BiLSTM layers, each with dimensions of 32, 30, 30, and 20. Each layer returns the complete sequence data and applies 0.1 dropout independently. The output is subjected to the Channel Attention mechanism to enhance feature importance and is subsequently flattened into a one-dimensional vector. A fully connected layer with 32 neurons and ReLU activation is used, followed by a final fully connected layer with 10 neurons and softmax activation for classification output.

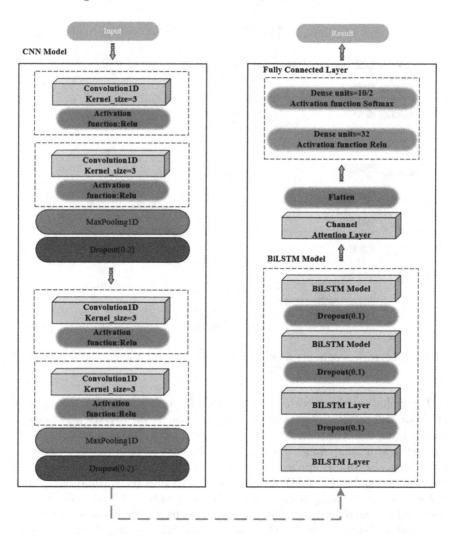

Fig. 3. The network architecture diagram

5   Comparative Experiments

To validate the effectiveness of our improved AGM under-sampling classification in conjunction with the CNN-BILSTM-ATTENTION model, in the binary classification experiments, we initially conducted controlled comparisons using different sampling techniques through the same CNN model and identical data preprocessing methods. Table 1 illustrates the impact of various sampling methods on the classification results.

Table 1. The metrics for different sampling methods processed through the same CNN model

Class Balanced	ACC(%)	DR(%)	FAR(%)	F1-score(%)
SCM	98.82	99.74	1.31	95.53
AGM	98.84	99.99	1.36	95.50
ROS	98.68	99.86	1.52	95.02
SMOTE	98.78	99.93	1.39	95.38
ADASYN	98.70	99.99	1.48	95.13
RUS+SM0TE	98.78	99.91	1.39	95.39
K-means+SMOTE	98.76	99.97	1.47	95.53

In the experiments, we employed representative metrics for imbalanced datasets, including Accuracy (ACC), False Alarm Rate (FAR), Detection Rate (DR), and F1-Score. These metrics are widely recognized as indicative of classification performance in imbalanced datasets. Their respective calculation formulas are provided in Eqs. 1 through 4, where TP denotes True Positives, TN stands for True Negatives, FP represents False Positives, and FN signifies False Negatives.

$$\text{Accuracy} = \frac{TP + TN}{TP + TN + FP + FN} \tag{1}$$

$$FAR = \frac{FP}{TP + FN} \tag{2}$$

$$DR = \frac{TP}{TP + FN} \tag{3}$$

$$F1 - Score = \frac{2 \cdot \text{Precision} \cdot \text{Recall}}{\text{Precision} + \text{Recall}} \tag{4}$$

According to the experimental results, the combination of ADASYN and GMM sampling demonstrates superior performance in terms of detection rate and accuracy compared to other sampling methods. Although there is a minor difference in false alarm rates and F1-Score in comparison to alternative sampling techniques, taking all factors into account, the modified AGM sampling method proves to be the most effective.

Figure 4 depicts the binary classification confusion matrix of the UNSW-NB15 dataset after SGM sampling followed by CNN processing, while Fig. 5 portrays the confusion matrix of the dataset after AGM sampling followed by CNN processing

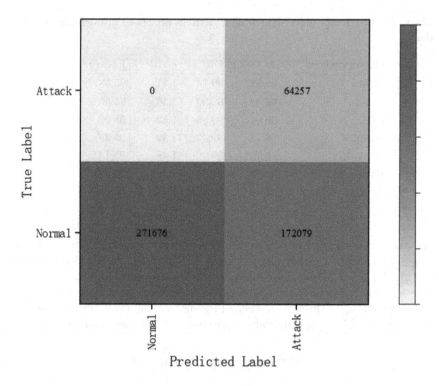

Fig. 4. SGM-Sampled UNSW-NB15

Subsequently, the data processed identically and balanced through AGM sampling for the UNSW-NB15 dataset is subjected to different models to assess their performance. Table 2 records the classification results after applying various models following the same data preprocessing and AGM balancing. Experimental results indicate that the improved CNN-BILSTM-ATT model exhibits notably higher F1-Scores in binary classification experiments. Although there is a marginal decrease in ACC and FAR metrics, the overall performance is enhanced compared to other models.

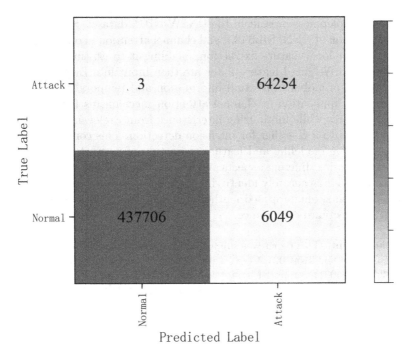

Fig. 5. AGM-Sampled UNSW-NB15

Table 2. Diverse Model Performance Metrics

Model	ACC(%)	DR(%)	FAR(%)	F1-score(%)
CNN	98.84	99.99	1.36	95.53
CNN-LSTM	98.80	99.99	1.37	95.48
CNN-BILSTM	98.80	99.99	1.37	95.49
CNN-GRU	98.78	99.99	1.37	95.48
CNN-BIGRU	98.80	99.99	1.37	95.48
CNN-BILSTM-ATT	98.80	99.99	1.37	95.82
CNN-BIGRU-ATT	98.80	99.99	1.37	95.48

6   Conclusion

This paper presents an enhanced hybrid sampling method that combines ADASYN with GMM to address the issue of data imbalance in network intrusion datasets. It effectively mitigates the problem of limited data samples for certain attack traffic categories while improving the quality of synthetic samples and better aligning them with the data distribution. Furthermore, we introduce a network data preprocessing method tailored to our proposed deep hierarchical network model, designed to tackle complex and multidimensional network threats. This preprocessing phase aims to eliminate noise, inconsistencies, and data incompleteness present in the UNSW-NB15 dataset. Subsequently, we enhance the fusion of CNN, BiLSTM, and channel attention in our approach. We employ CNN for local feature extraction, enabling it to capture local patterns within the data. Weighted feature maps are then input into BiLSTM, facilitating the capture of global contextual information and temporal relationships in sequences. The application of channel attention mechanisms further enhances important features while minimizing interference from irrelevant ones. The output is connected to a classifier for intrusion detection. This combination allows for comprehensive modeling and feature extraction of network intrusion detection data, addressing different aspects and levels of the problem. Consequently, it assists the model in accurately identifying malicious activities. In binary classification experiments, our proposed method demonstrates a notable improvement in intrusion detection performance.

Acknowledgment. This paper was supported by the National Natural Science Foundation of China (No.92159102), the Natural Science Foundation of Jiangxi Province (No.20232ACB205001), Support Plan for Talents in Gan Poyang – Academic and Technical Leader Training Project in Major Disciplines (No.20232BCJ22025).

References

1. Tesfahun, A., Bhaskari, D.L.: Intrusion detection using random forests classifier with smote and feature reduction. In: 2013 International Conference on Cloud & Ubiquitous Computing & Emerging Technologies, pp. 127–132. IEEE (2013)
2. Binbusayyis, A., Vaiyapuri, T.: Identifying and benchmarking key features for cyber intrusion detection: an ensemble approach. IEEE Access **7**, 106495–106513 (2019)
3. Koroniotis, N., Moustafa, N., Sitnikova, E., Slay, J.: Towards developing network forensic mechanism for botnet activities in the IoT based on machine learning techniques. In: Hu, J., Khalil, I., Tari, Z., Wen, S. (eds.) MONAMI 2017. LNICST, vol. 235, pp. 30–44. Springer, Cham (2018). https://doi.org/10.1007/978-3-319-90775-8_3
4. Yin, C., Zhu, Y., Fei, J., He, X.: A deep learning approach for intrusion detection using recurrent neural networks. IEEE Access **5**, 21954–21961 (2017)
5. Yong Zhang, X., Chen, D.G., Song, M., Teng, Y., Wang, X.: PCCN: parallel cross convolutional neural network for abnormal network traffic flows detection in multiclass imbalanced network traffic flows. IEEE Access **7**, 119904–119916 (2019)
6. Jiang, K., Wang, W., Wang, A., Haibin, W.: Network intrusion detection combined hybrid sampling with deep hierarchical network. IEEE access **8**, 32464–32476 (2020)
7. Lo, W.W., Layeghy, S., Sarhan, M., Gallagher, M., Portmann, M.: E-GraphSAGE: a graph neural network based intrusion detection system for IoT. In: NOMS 2022-2022 IEEE/IFIP Network Operations and Management Symposium, pp. 1–9. IEEE (2022)
8. Yanfang, F., Yishuai, D., Cao, Z., Li, Q., Xiang, W.: A deep learning model for network intrusion detection with imbalanced data. Electronics **11**(6), 898 (2022)
9. Moustafa, N., Slay, J.: UNSW-NB15: a comprehensive data set for network intrusion detection systems (UNSW-NB15 network data set). In: 2015 Military Communications and Information Systems Conference (MilCIS), pp. 1–6. IEEE (2015)

10. Zhang, H., Wu, C.Q., Gao, S., Wang, Z., Xu, Y., Liu, Y.: An effective deep learning based scheme for network intrusion detection. In: 2018 24th International Conference on Pattern Recognition (ICPR), pp. 682–687. IEEE (2018)
11. Zhang, H., Huang, L., Wu, C.Q., Li, Z.: An effective convolutional neural network based on smote and gaussian mixture model for intrusion detection in imbalanced dataset. Comput. Netw. **177**, 107315 (2020)
12. Chawla, N.V., Bowyer, K.W., Hall, L.O., Kegelmeyer, W.P.: SMOTE: synthetic minority over-sampling technique. J. Artif. Intell. Res. **16**, 321–357 (2002)
13. He, H., Bai, Y., Garcia, E.A., Li, S.: ADASYN: adaptive synthetic sampling approach for imbalanced learning. In: 2008 IEEE International Joint Conference on Neural Networks (IEEE World Congress on Computational Intelligence), pp. 1322–1328. IEEE (2008)

Using the SGE-CGAM Method
to Address Class Imbalance Issues
in Network Intrusion Detection

Xin Chen[(✉)] [ID], Ke Yi [ID], and Jia Yu [ID]

East China Jiaotong University, Nanchang, China
3137551149@qq.com, 2612@ecjtu.edu.cn

Abstract. Network Intrusion Detection Systems (NIDS) play a crucial role in safeguarding network security and data integrity. Nevertheless, the challenge of class imbalance within intrusion detection datasets
hampers the classifier's performance on minority classes. To simultaneously improve detection precision while maintaining efficiency, we introduce an innovative approach for addressing class imbalance in extensive datasets, denoted as SGE (SMOTE-Gaussian Mixture Model-Edited
Nearest Neighbors). This approach amalgamates oversampling techniques, such as SMOTE, with the Gaussian Mixture Model (GMM), and
undersampling techniques, including Edited Nearest Neighbors (ENN).
Additionally, we have enhanced a deep learning model architecture,
denoted as the CNN-biGRU-ATT Model (Convolutional Neural Network
- bidirectional Gated Recurrent Unit - Attention mechanism), which
integrates Convolutional Neural Networks (CNN), bidirectional Gated
Recurrent Units (biGRU), and attention mechanisms (ATT). We ultimately devised a hybrid network intrusion detection model, SGE-CGAM,
which combines class imbalance handling with a blended deep learning
model. We validated the superiority of this model using the CICIDS2017
dataset. The results obtained from the experiments demonstrate that
for binary-class and multi-class tasks on the CICIDS2017 dataset, SGE-
CGAM achieved Precision rates of 99.98% and 99.76%, respectively,
underscoring SGE-CGAM as an effective intrusion detection solution.

Keywords: NIDS · Class Imbalance Handling · SGE · Deep Learning
Architectures for Intrusion Detection · CICIDS2017 Dataset

1 Introduction

As 5G technology becomes increasingly prevalent and 6G technology continues to
evolve, computer technology has greatly facilitated human production and daily
life. However, it has also led to the generation of massive amounts of data on the

Supported by organizations of the National Natural Science Foundation of China, the
Natural Science Foundation of Jiangxi Provincial, and the Major Discipline Academic
and Technical Leaders Training Program of Jiangxi Province.

J. Vaidya et al. (Eds.): AIS&P 2023, LNCS 14510, pp. 176–186, 2024.
https://doi.org/10.1007/978-981-99-9788-6_15

network, posing challenges in terms of storage and transmission [1]. Individuals, organizations, and nations rely on networks for the storage and transmission of sensitive information, and any vulnerabilities or threats can impact the entire network [2]. This underscores the critical importance of network security. Clearly, network intrusion detection is an exceedingly challenging task, characterized by various complexities and difficulties, such as the vast volume of data, class imbalance, and limitations of traditional defense methods [3]. However, due to the increasing variety of attack types and the exponential surge in network traffic, conventional machine learning is no longer sufficient to fulfill the requirements of large-scale network intrusion defense.

The objective of this study is to explore approaches for tackling the mentioned challenges. In this regard, we introduce a novel hybrid network intrusion detection model, SGE-CGAM. The SGE-CGAM model integrates the SGE (SMOTE-Gaussian Mixture Model-Edited Nearest Neighbors) technique for class imbalance handling with the CGAM (CNN-biGRU-ATT) deep learning architecture, which combines the strengths of convolutional neural networks (CNN), bidirectional gated recurrent units (biGRU), and attention mechanisms (ATT). We evaluate this model in both binary and multi-class scenarios using the CICIDS2017 dataset.

2   Proposed Solution

The hybrid network intrusion detection model suggested in this study, SGE-CGAM, is depicted in Fig. 1 and comprises the integration of the class imbalance handling technique SGE with the CGAM NIDS architecture. The hybrid model comprises three primary components: the data preprocessing module, the class imbalance handling module, and the model training module. To assess its performance in contemporary network environments, we conducted binary and multi-class classification experiments on the CICIDS2017 dataset as detailed below.

2.1   Raw Dataset Description

The dataset employed in our study is the CICIDS2017 dataset [4], which was meticulously curated and compiled by the Canadian Cyber Security Institute in collaboration with the B-Profile system [5] at the close of 2017. This dataset was created within a simulated real-world network environment with the aim of closely emulating network traffic and attacks as they manifest in practical scenarios. It encompasses a total of 2,830,473 network traffic samples, encompassing one benign class and fourteen distinct attack classes. The attack categories encompass the most prevalent attack types, and the comprehensive data distribution for each class is displayed in Table 1.

This dataset features the extraction of 84 features from generated network traffic, with the final column representing multi-class labels [6]. Moreover, in contrast to other publicly available datasets, the CICIDS2017 dataset possesses

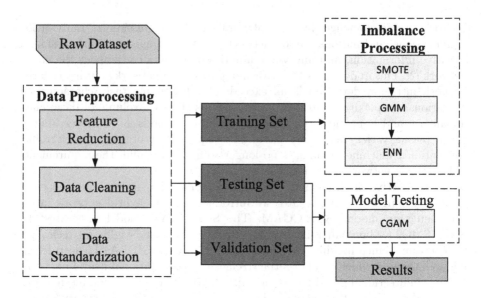

Fig. 1. A schematic diagram of the SGE-CGAM NIDS model.

a significant advantage in terms of its capacity to reflect modern network environments and diverse attack techniques. It fully complies with all eleven criteria for performance evaluation [7]. In our experiments, we utilized all samples from the CICIDS2017 dataset and partitioned it into training, validation, and testing sets, with a ratio of 7:1:2.

2.2   Data Preprocessing

For the CICIDS2017 dataset, the data preprocessing module encompasses three steps: Feature Reduction, Data Cleaning, and Data Standardization.

Firstly, in the Feature Reduction step, we removed unique identifiers such as "Flow ID", "Source IP", "Source Port", "Destination IP", and "Protocol" that do not provide direct information about traffic or anomalies. Additionally, to reduce input dimensionality and computational complexity, we further eliminated a continuous timestamp, "Time stamp". This reduced the feature dimension to 78. Next, in the Data Cleaning step, the CICIDS2017 dataset contained values labeled as "infinity". We replaced these "infinity" values with the largest finite value and also substituted missing values (NaN) with 0. Finally, in the Data Standardization step, as per Eq. (1), we standardized all remaining features and normalized them to follow a Gaussian distribution with a mean of 0 and a standard deviation of 1:

$$\mathbf{x}' = \frac{x - \mu}{\sigma} \tag{1}$$

Table 1. The sample count within each class of the CICIDS2017 dataset

Class	Categorical	Binary
BENIGN	2273097	2273097
DoS Hulk	231073	557646
PortScan	158903	
DDoS	128027	
DoS GoldenEye	10293	
FTP-Patator	7898	
SSH-Patator	5897	
DoS slowloris	5796	
DoS Slowhttptest	5499	
Bot	1966	
Web Attack Brute Force	1507	
Web Attack XSS	652	
Infiltration	36	
Web Attack Sql Injection	21	
Heartbleed	11	

where x represents the original features, $\mathbf{x}'$ denotes the normalized features, and μ and σ correspond to the mean and standard deviation of the features, respectively. The normalized data maintains the same linear relationship as the original data [8]. Data normalization aids in enhancing the stability, training speed, and performance of the model, while also rendering the data more manageable and interpretable.

2.3   Class Imbalance Processing

As shown in Table 1, it is evident that a portion of the CICIDS2017 dataset exhibits imbalances in both multi-class and binary labels [9]. Moreover, within this dataset, the number of anomalous samples is inherently scarce in comparison to normal samples. For instance, the "Infiltration" class in the CICIDS2017 dataset contains only 36 samples, accounting for a mere 0.0013% of the entire dataset. Consequently, standalone undersampling or oversampling techniques are inappropriate. In this paper, we introduce an innovative SGE approach, which combines SMOTE and ENN while incorporating GMM. This method facilitates the resampling of samples from all classes to a unified quantity, thereby achieving a more balanced distribution of sample counts across different classes.

SMOTE is a classic oversampling technique introduced by Chawla et al. in 2002 [10]. It is a method employed to address class imbalance issues by augmenting the number of minority class samples through the generation of synthetic samples. The fundamental concept behind SMOTE involves randomly selecting a sample from the minority class and then randomly selecting one or more samples from its nearest neighbors to generate new synthetic samples through linear interpolation.

The Gaussian Mixture Model (GMM) is a statistical model used for modeling the distribution of multidimensional data. It assumes that data is composed of a mixture of several Gaussian distributions, with each Gaussian distribution representing a cluster within the data. GMM fits the data distribution by estimating parameters such as mean, covariance matrix, and weights for each distribution. GMM stands as a robust tool for probabilistically modeling data attributes. Especially pertinent in data balancing tasks, the GMM adeptly discerns and encapsulates various patterns and nuances. From a mathematical perspective, the GMM can be delineated as presented in Eq. (2):

$$p(x) = \sum_{k=1}^{K} \pi_k \mathcal{N}(x|\mu_k, \Sigma_k) \tag{2}$$

where $p(x)$ represents the probability density of data point for a given data point x. The model encompasses a mixture of k Gaussian distributions, where each distribution is defined by its mixture coefficient π_k, mean μ_k, and covariance matrix Σ_k.

Edited Nearest Neighbors (ENN), on the other hand, is a data cleaning method primarily employed to remove samples from a dataset with minimal influence on classifier performance in classification tasks. By analyzing the nearest neighbor relationships of each sample, ENN identifies and eliminates data points that may introduce noise or redundant information into the dataset, thus enhancing the stability and generalization capability of the model.

The SGE method aims to balance data, improve data quality, and enhance model performance simultaneously. Initially, we employ the SMOTE algorithm to generate synthetic minority class samples, ensuring an augmented representation of the minority class. Subsequently, we apply GMM clustering to the entire dataset, dividing it into distinct components or clusters. This aids in understanding the data distribution and structure. Then, we utilize ENN to remove noisy data from the majority class that has minimal impact on the classifier. Finally, we merge the enhanced and cleaned data with the original dataset, resulting in a balanced and cleaner dataset.

2.4   Model Testing

The CNN architecture is more effective in extracting spatial features from data flows, while its ability to capture long-distance dependency information is moderate. The convolution operation can be mathematically represented as depicted in Eq. (3):

$$y(t) = \sum_{i=0}^{M-1} w(i) \cdot x(t+i) + b \tag{3}$$

where $y(t)$ denotes the output at time t, w is the kernel or filter with a length of M, b refers to the bias term, the convolutional operation involves sliding the kernel w across the input sequence x and computing the dot product at each position.

On the other hand, biGRU is a recurrent neural network capable of effectively extracting remote dependency information and capturing context bidirectionally, albeit with a higher parameter count and longer training time. Combining these two components enhances the model's feature learning capabilities, allowing for comprehensive feature extraction from both spatial and temporal dimensions, ultimately leading to improved classification and detection accuracy. The introduction of the attention mechanism (ATT) into the model aims primarily to enhance the model's representation capacity when dealing with complex data and tasks, the equation for the attention weights is presented as Eq. (4):

$$alpha_t = \frac{\exp(e_t)}{\sum_{i=1}^{T} \exp(e_i)} \tag{4}$$

The score e_t can be expressed as per Eq. (5):

$$e_t = a(s_{t-1}, h_t) \tag{5}$$

where s_{t-1} represents the hidden state of the decoder at the previous time step, h_t denotes the hidden state of the encoder at time t, the function a can be a feedforward network or a simple dot product, contingent upon the attention architecture.

Within the model training module, we have designed an improved deep learning model structure known as CGAM (CNN-biGRU-ATT Model). It primarily consists of three components: firstly, spatial features are extracted through a four-layer one-dimensional CNN module; secondly, temporal features are extracted through two layers of bi-GRU; next, the attention mechanism is employed to enhance the model's focus on different parts of the input data. Finally, the Softmax function is applied to the feature outputs for the ultimate classification. The specific structure is depicted in Fig. 2.

3   Experimental Analysis

In the context of our study, the CICIDS2017 dataset was employed, undergoing preprocessing to align with our fusion model that integrates 1D CNN, Bidirectional GRU, and an attention mechanism. Hyperparameters were chosen empirically, setting the learning rate to 0.005-subject to dynamic adjustments contingent on validation loss-with a batch size of 256, facilitated by the NAdam optimizer. Throughout the training phase, model weights were periodically saved after each epoch. The model's performance was meticulously evaluated against metrics such as accuracy, precision, recall, and F1 score, underscoring a holistic assessment of our intrusion detection system's capabilities.

3.1   Evaluation Metrics

We employed the conventional four evaluation metrics to assess the performance of the SGE-CGAM architecture. These four metrics are accuracy (ACC), precision, recall, and F1-score. These evaluation metrics are essential for gauging

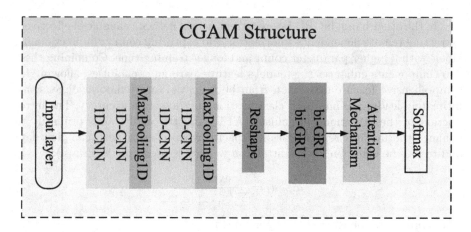

Fig. 2. CNN-biGRU-ATT Model structure diagram.

the performance of classification models. They are computed by contrasting the model's predictions with the actual ground truth. The formulations for these metrics are delineated in Eqs. (6)-(9):

$$ACC = \frac{TP + TN}{TP + TN + FP + FN} \tag{6}$$

$$Presision = \frac{TP}{TP + FP} \tag{7}$$

$$\mathrm{Re}\,call = \frac{TP}{TP + FN} \tag{8}$$

$$F1 - Score = \frac{2 * \mathrm{Precision} * \mathrm{Recall}}{\mathrm{Precision} + \mathrm{Recall}} \tag{9}$$

where TP (True Positives) denotes true positive instances, TN (True Negatives) represents true negative instances, FP (False Positives) signifies false positive instances, and FN (False Negatives) indicates false negative instances.

3.2   Binary-Class and Multi-class Classifications

To evaluate the performance of the proposed SGE-CGAM architecture for network intrusion detection, a sequence of experiments was carried out. Initially, the imbalance handling method, SMOTE+GMM+ENN (SGE), was validated utilizing the model introduced in this paper. Binary classification was employed to compare different imbalance handling methods, and the experimental outcomes are displayed in Table 2. Subsequently, the CNN-biGRU-ATT (CGAM) model was validated using the same imbalance handling method. Four models, namely CNN, GRU, RNN, and CNN-biGRU-ATT, were compared, and the results of the experiments are summarized in Table 3.

Table 2. Performance Metrics for Binary-Class Classification on the CICIDS2017 Dataset with Different Imbalance Handling Methods (%)

Dataset	Imbalance Processing Method	Accuracy	Precision	Recall	F1-Score
CICIDS2017	STMOE+GMM	99.68	99.71	99.68	99.65
	STMOE+GMM+RENN	99.57	99.59	99.57	99.55
	STMOE+GMM+ENN	**99.88**	**99.89**	**99.88**	**99.86**

Table 3. Performance Metrics (%) for Binary-Class Classification of CNN, GRU, RNN, and CNN-biGRU-ATT on the CICIDS2017 Dataset

Dataset	Model	Accuracy	Precision	Recall	F1-Score
CICIDS2017	CNN	98.74	98.80	98.74	98.76
	GRU	98.95	98.58	98.95	98.45
	RNN	91.68	93.05	91.68	92.03
	CNN-biGRU-ATT	**99.88**	**99.89**	**99.88**	**99.86**

From Tables 2 and 3, it is evident that on the open-source CICIDS2017 dataset, the imbalance handling method, SGE, and the CGAM model yield the most favorable results. Furthermore, during the experimentation process, we obtained the confusion matrix for SGE-CGAM, as depicted in Fig. 3, confirming the architecture's strong performance in binary classification.

To further assess the effectiveness of the proposed architecture in network intrusion detection, we conducted multi-class classification with SGE-CGAM on the CICIDS2017 dataset, as illustrated in Table 4 and Fig. 4. The experimental results demonstrate that SGE-CGAM also exhibits commendable performance in multi-class classification.

Table 4. Performance Metrics (%) for Multi-Class Classification of SGE-CGAM on the CICIDS2017 Dataset

Dataset	Method	Accuracy	Precision	Recall	F1-Score
CICIDS2017	SGE-CGAM	**99.63**	**99.76**	**99.63**	**99.68**

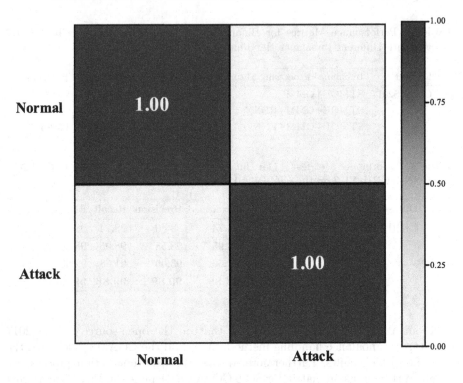

Fig. 3. Confusion Matrix for Binary-Class Classification of SGE-CGAM.

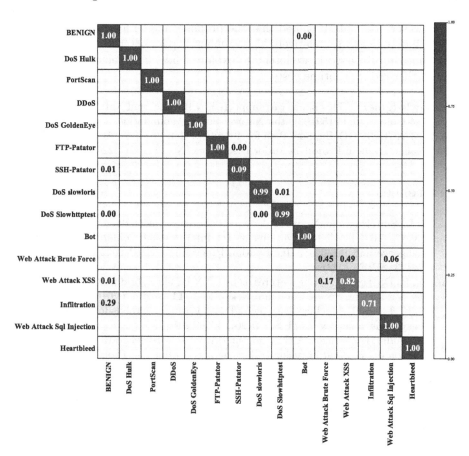

Fig. 4. Confusion Matrix for Multi-Class Classification of SGE-CGAM.

4   Conclusion

This paper integrates the imbalance handling technique SGE with the CGAM model, proposing a hybrid network intrusion detection model: SGE-CGAM. SGE effectively addresses the issue of imbalanced sample labels in network intrusion detection datasets, while comprehensive feature learning is achieved through the fusion of CNN and biGRU. Additionally, the introduction of an attention module enhances the model's capability to represent complex data, thus improving its performance. The SGE-CGAM model achieved a Precision of 99.98% for binary-class classification and 99.76% for multi-class classification on the CICIDS2017 dataset, providing substantial evidence for the effectiveness of the SGE-CGAM model in network intrusion detection on large-scale imbalanced datasets.

However, it should be noted that the proposed model has some limitations, primarily related to its relatively high number of parameters and longer execution time. In future work, we intend to explore model lightweighting techniques and further reduce runtime costs.

Acknowledgment. This paper was supported by the National Natural Science Foundation of China (No.92159102), the Natural Science Foundation of Jiangxi Province (No.20232ACB205001), Support Plan for Talents in Gan Poyang – Academic and Technical Leader Training Project in Major Disciplines (No.20232BCJ22025).

References

1. Gu, L., Sun, H., Zhao, X., Wang, L.: A distributed intrusion detection system based on CNN-GRU in cloud environment. In: International Conference on Artificial Intelligence and Intelligent Information Processing (AIIIP 2022), vol. 12456, pp. 167–173. SPIE (2022)
2. Sun, P.: Dl-IDS: extracting features using CNN-LSTM hybrid network for intrusion detection system. Secur. Commun. Networks 1–11, 2020 (2020)
3. Liu, H., Lang, B.: Machine learning and deep learning methods for intrusion detection systems: a survey. Appl. Sci. 9(20), 4396 (2019)
4. Sharafaldin, I., Lashkari, A.H., Ghorbani, A.A.: Toward generating a new intrusion detection dataset and intrusion traffic characterization. In: Proceedings of the 4th International Conference on Information Systems Security and Privacy (ICISSP 2018), pp. 108–116 (2018)
5. Sharafaldin, I., Gharib, A., Lashkari, A.H., Ghorbani, A.A.: Towards a reliable intrusion detection benchmark dataset. Softw. Networking 2017(1), 177–200 (2017)
6. Gharib, A., Sharafaldin, I., Lashkari, A.H., Ghorbani, A.A.: An evaluation framework for intrusion detection dataset. In: 2016 International Conference on Information Science and Security (ICISS), pp. 1–6. IEEE (2016)
7. Abdulhammed, R., Musafer, H., Alessa, A., Faezipour, M., Abuzneid, A.: Features dimensionality reduction approaches for machine learning based network intrusion detection. Electronics 8(3), 322 (2019)
8. Xiao, Y., Xing, C., Zhang, T., Zhao, Z.: An intrusion detection model based on feature reduction and convolutional neural networks. IEEE Access 7, 42210–42219 (2019)
9. Čavojský, M., Bugár, G.,šan Levický, D.: Comparative analysis of feed-forward and RNN models for intrusion detection in data network security with UNSW-NB15 dataset. In: 2023 33rd International Conference Radioelektronika (RADIOELEK-TRONIKA), pp. 1–6. IEEE (2023)
10. Chawla, N.V., Bowyer, K.W., Hall, L.O., Kegelmeyer, W.P.: SMOTE: synthetic minority over-sampling technique. J. Artif. Intell. Rese. 16, 321–357 (2002)

A Study of Adaptive Algorithm for Dynamic Adjustment of Transmission Power and Contention Window

Qi Shi[✉] , BoTao Tu , and Guolong Zhang

East China Jiaotong University, Nanchang, China
907044792@qq.com

Abstract. Vehicular Ad-Hoc Networks (VANETs) enable information sharing among vehicles, enhancing the safety of vehicle operations and holding significant practical value in preventing and managing events such as traffic congestion and accidents. However, effectively distributing beacon messages in complex and dynamic traffic environments presents a major challenge. Therefore, this paper introduces VANETs, where the strategy for communication link duration involves improving transmission range by increasing transmission power. However, under dense traffic conditions, boosting transmission power may lead to high interference levels and increased network overhead. Consequently, dynamic adjustment of power based on varying traffic density has become a common strategy.

Keywords: Vehicular Ad Hoc Network (VANET) · Beacon Messages · Dynamic Adjustment · varying traffic density

1 Introduction

This article investigates the beacon message transmission mechanism and designs two joint adaptive control schemes for urban road environments where parameters such as broadcast cycle and transmission power cannot adaptively adjust. Considering the highly predictable vehicle node information in beacon messages, traffic flow prediction is performed using an LSTM model to obtain traffic flow prediction parameters. These predicted parameters are then used to assess the channel status, enabling adaptive joint control of beacon message broadcast cycles and transmission power.

Additionally, another approach involves estimating vehicle density and designing a transmission power adaptation method. An analysis of the relationship between the contention window and the number of idle slots is conducted to

This paper was supported by the National Natural Science Foundation of China (No. 92159102), the Natural Science Foundation of Jiangxi Province (No. 20232ACB205001), Support Plan for Talents in Gan Poyang – Academic and Technical Leader Training Project in Major Disciplines (No. 20232BCJ22025).

design a contention window adaptation method. Finally, an adaptive scheme for beacon message transmission power and contention window is proposed. Both of these approaches are simulated and analyzed, and the results show improvements in beacon message transmission performance.

Vehicular Ad Hoc Networks (VANETs) play a crucial role in improving traffic congestion and reducing traffic accidents. Traffic state perception largely relies on the beacon message transmission mechanism. This article focuses on the research and analysis of the beacon message transmission mechanism in urban road environments within vehicular self-organizing networks.

The IEEE 802.11p criteria of VANET [1–3] utilizes Enhanced Distributed Channel(EDCA) Access from the 802.11e criteria [4] to support quality of service for different applications. According to the 802.11e criteria, all kinds of priorities will be allocated to vehicles based on the importance of traffic-related messages, However, The current EDCA cannot provide predictive services for the current IEEE 802.11p MAC. Beacon messages are broadcasted over the shared channel, and When multiple nodes are located within a communication scope attempt to broadcast at the same time, it can result in a high probability of transmission conflicts. Furthermore, without collision detection frames, CW size does not double when beacon messages collide. Researchers expect the CW value to be sufficiently large so that the probability of multiple nodes back off the same number of slots is small enough. But CW too large can also lead to the increased of latency. In summary, dynamic adjustment of beacon message CW size based on changes in vehicle density is essential. Based on the above analysis, this section proposes This paper proposes a more comprehensive way to balance the size of the CW value.

Joint Adaptive Control of Transmission Power and Contention Window: Vehicles dynamically select suitable transmission power based on locally estimated density. Adaptive power adjustments are made for each vehicle. Additionally, the article analyzes the relationship between idle slots and the optimal contention window value using a mathematical model. The adaptive adjustment of the contention window is achieved by comparing observed values with thresholds. Simulation analyses reveal significant improvements in network performance metrics such as throughput and delay.

2   Materials and Methods

The principle of DCF [5,6] is based on a contention mechanism, providing distributed access where multiple distributed wireless nodes contend for the same resource. DCF uses carrier sensing [7] and collision avoidance techniques to prevent collisions on a single shared channel, while channel access for distributed data traffic is completed through the RTS/CTS mechanism.

2.1   Interframe Spacing

The interval between frames is referred to as interframe spacing (IFS). DCF protocol includes Short Interframe Space (SIFS), Distributed Coordination Function

Interframe Space (DIFS), and Extended Interframe Space (EIFS), each different lengths of IFS. SIFS is the shortest and is used for responding to CTS frames or ACK frames. DIFS is unique to DCF agreement, in situations where nodes are required to continuously monitor the channel to detect its idle state before initiating data transmission. If the channel is found to be idle, the node is mandated to observe a DIFS (Distributed Inter-Frame Space) period before commencing data transmission. However, if the channel is perceived as busy at any moment during the DIFS interval, the transmission is postponed. EIFS is the longest and is used for backoff in case a node cannot decode a data packet, ensuring successful transmission.

2.2   CSMA/CA Technique

To avoid collisions caused by multiple nodes trying to enter the network meanwhile, DCF protocol employs a random access mechanism known as Carrier Sense Multiple Access with Collision Avoidance (CSMA/CA) [8]. According to CSMA/CA, before transmitting data, nodes first listen to the channel's status. If the channel is idle, the node waits for a DIFS interval before sending data and then waits for acknowledgment. After the recipient node effectively acquires the data, it waits for an SIFS interval before sending an acknowledgment frame (ACK). The sending node acknowledges successful transmission upon receiving the ACK frame. If the channel is busy, the node delays access and enters a backoff stage based on the backoff time after detecting specific frames.

2.3   RTS/CTS Mode

DCF has double channel modes: the basic access mode and the RTS/CTS access mode [9,10]. The basic involves sender detecting a continuously idle channel for a DIFS interval, then sending data to the destination node. Upon successful reception, the sender waits for an SIFS interval to receive an ACK frame. If there is no timeout, the transmission is confirmed as successful; otherwise, the sender enters a backoff stage for retransmission until success or the maximum retransmission limit is reached, at which point the data is discarded. The sender detects a channel idle for a DIFS interval and sends an RTS frame to the target node. If the RTS frame is received, the target node waits for an SIFS interval and responds with a CTS frame. Data transmission can then occur. Upon receiving the CTS frame without exceeding the timeout, the sender waits for an SIFS interval and sends data. The subsequent process is similar to the basic access mode. During this process, other nodes update their Network Allocation Vector values to defer access based on the RTS and CTS frames. If the sending node sends an RTS frame and does not receive a CTS frame, it enters a backoff stage and retransmits the RTS frame until successful transmission or reaching the maximum retransmission limit, at which point the data is lost. The DCF protocol employs a binary exponential backoff algorithm, and the random backoff time T is selected according to the following formula.

$$T = INT(CW_i \times Random()) \times SlotTime \tag{1}$$

3 Experimental Results

In VANETs, the transmission power is typically fixed, which means that the transmission range remains constant. This is clearly not suitable for vehicle nodes in rapidly changing topological structures. Using high transmission power implies a larger transmission range, which is beneficial for the perception of vehicle nodes. However, it can also interfere with the transmissions of neighboring nodes. Low transmission power, on the other hand, represents a smaller transmission range, significantly reducing perception capabilities and making it more susceptible to environmental interference. Based on these considerations, this paper presents an adaptive transmission power method based on vehicle density estimation, considering whether to increase or decrease power based on the estimated vehicle density. Firstly, vehicle density is estimated. In vehicular ad-hoc networks, there are generally two methods to accept the feedback online. The proactive method involves nodes monitoring the network and exchanging information with neighboring nodes. To illustrate, a node can exchange its list of all exchange information about their immediate neighbors with nearby nodes. The disadvantage of the proactive approach is the elevated network overhead, leading to additional bandwidth consumption. In VANETs, bandwidth is a scarce resource, so reducing extra bandwidth consumption. Proactive method augment the information to sent causing extra congestion. Secondly, passive methods could be employed to accept the feedback. In Mobile Network, Nodes within their communication range actively monitor all transmitted messages. For example, if all nodes have a transmission range of 300 m, a node will listen to all messages sent within a 300-m range. Nodes can note statistical information about network conditions based on the received packets. Passive methods need not extra website input. Therefore, this paper uses the passive approach to receive network feedback. After obtaining feedback, vehicle density is estimated.

$$K = \frac{N_i}{(D_f + D_b) \cdot N_l} \tag{2}$$

N_i represents the quantity detected by current nodes, D_f, D_b represents the distance to the farthest leading and trailing vehicles estimated by the vehicle node, and N_1 represents the number of lanes. The expression for the transmission range (TR) can be derived as follows:

$$TR = min\{L(1 - K), \sqrt{\frac{LInL}{K}} + aL\} \tag{3}$$

a represent the traffic flow, L represent the length of the road, and p represent estimated car density. Because the relationship between transmission power and transmission distance corresponds to an empirical value, and the complex and variable nature of the vehicular environment makes it challenging to calculate directly, researchers conduct simulation studies of VANETs in various scenarios. Table 1 illustrates the relationship between transmission power and distance.

Table 1. Beacon Message Transmission Distance and Transmission Power Mapping Table

transmission distance	transmit power	transmission distance	transmit power
0–9	−20	380–449	14
10–49	−12	450–549	17
50–100	−5	550–649	20
100–125	−3	650–749	24
126–149	1	750–849	27
150–209	4	850–929	29
210–299	6	939–970	31
300–349	10	971–1000	32

At the outset, individual vehicles commenced with random transmission power settings and monitored information from other vehicles. To alleviate the drawbacks associated with excessive transmission power and enhance the longevity of communication links in scenarios characterized by sparse vehicle traffic, each vehicle dynamically adapts its transmission power based on the estimated local vehicle density. Formula (2) is employed to estimate the vehicle density within its transmission range. Leveraging the estimated vehicle density, the algorithm computes the transmission range using Formula (3) and subsequently configures the corresponding transmission power, based on a lookup table. The specific procedure is as follows: "At the outset, individual vehicles initiate communication at arbitrary transmission power levels while listening to information from other vehicles. To alleviate the detrimental impacts of elevated transmission power and prolong communication links in scenarios characterized by sparse vehicle traffic, each vehicle dynamically modulates its transmission power in accordance with the estimated local vehicle density. Formula (2) is employed to estimate the local car density within its transmission coverage area. Leveraging the estimated vehicle density, the algorithm utilizes Formula (3) to calculate the transmission range, and then configures the corresponding transmission power accordingly based on a lookup table. B. Experimental Simulation and Results Analysis To validate the algorithm's performance, experiments were conducted using the NS2.35 network simulation software to analyze the adaptive algorithms for transmission power and contention window. Table 2 represents the parameters used in this experiment (Fig. 1):

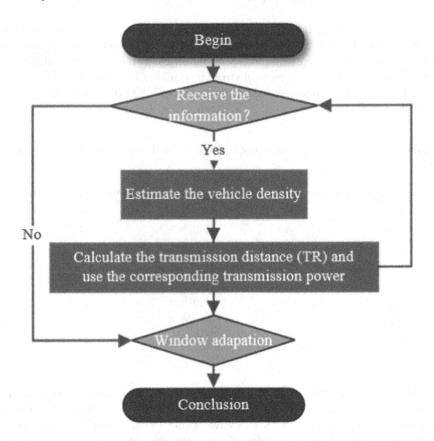

Fig. 1. Overall structure flow chart

Table 2. Configuration of Experimental Parameters

parameter	value
bandwidth	5.89 GHz
channel model	Two-ray ground model
contention window	[15, 1023]
vehicle density	0–100
packet size	512 Byte
SIFS	32 μs
DIFS	50 μs
$1/\lambda_1$	1.2
λ_2	0.0005

Figure 2 illustrates the variation of packet reception rate with vehicle density. In comparison to the conventional 802.11p program, both LIMERIC algorithm and our algorithm adaptively control beacon parameters, resulting in higher packet reception rates for both. Although the LIMERIC algorithm adjusts the message rate to some extent, alleviating beacon message congestion, our joint control approach can adaptively control both transmission power and contention window, effectively mitigating congestion in vehicle nodes. From the results, it is evident that our algorithm outperforms the LIMERIC algorithm and the traditional 802.11p scheme.

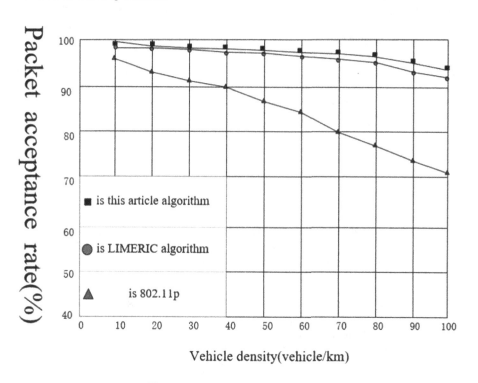

Fig. 2. Overall structure flow chart

4   Conclusion

This chapter primarily introduces a beacon message broadcast transmission scheme that VANETs using joint adaptive control of transmission power and contention window (CW) size. This method allows vehicles to dynamically select appropriate transmission power based on estimated local density. It combines the monitoring of idle slots to assess channel conditions, adaptively adjusting contention window size by comparing observed values to thresholds. Through simulation experiments, it is evident that the approach presented in this section outperforms traditional methods, resulting in improved throughput, higher packet reception rates, and reduced latency.

It employs predicted traffic flow parameters to adaptively control the beacon message broadcast cycle and transmission power. Simulation results demonstrate that this approach effectively enhances data packet reception rates, reduces latency, and lowers channel busyness.

Given the significant impact of MAC layer parameters on beacon message transmission performance, this paper considers cross-layer joint control of physical layer and MAC layer parameters for beacon messages. In the transmission power and contention window adaptation scheme, it initially adjusts transmission power based on vehicle density estimation. Subsequently, it dynamically adapts contention window values by observing the number of idle slots. Simulation results indicate that this approach can improve system throughput and reduce end-to-end latency.

Acknowledgements. This paper was supported by the National Natural Science Foundation of China (No. 92159102), the Natural Science Foundation of Jiangxi Province (No. 20232ACB205001), Support Plan for Talents in Gan Poyang – Academic and Technical Leader Training Project in Major Disciplines (No. 20232BCJ22025).

References

1. Araujo, G.B., Peixoto, M.L., Sampaio, L.N.: NDN4IVC: a framework for simulating and testing of applications in vehicular named data networking. arXiv preprint arXiv:2107.00715 (2021)
2. Kamel, J., Wolf, M., Van Der Hei, R.W., Kaiser, A., Urien, P., Kargl, F.: VeReMi extension: a dataset for comparable evaluation of misbehavior detection in VANETs. In: 2020 IEEE International Conference on Communications (ICC), ICC 2020, pp. 1–6. IEEE (2020)
3. Al Mallah, R., Quintero, A., Farooq, B.: Distributed classification of urban congestion using VANET. IEEE Trans. Intell. Transp. Syst. **18**(9), 2435–2442 (2017)
4. Eckhoff, D., Sommer, C.: A multi-channel IEEE 1609.4 and 802.11p EDCA model for the veins framework. In: Proceedings of 5th ACM/ICST International Conference on Simulation Tools and Techniques for Communications, Networks and Systems: 5th ACM/ICST International Workshop on OMNet++, Desenzano, Italy, 19–23 March 2012. OMNeT (2012)
5. Lukezic, A., Vojir, T., Cehovin Zajc, L., Matas, J., Kristan, M.: Discriminative correlation filter with channel and spatial reliability. In: Proceedings of the IEEE Conference on Computer Vision and Pattern Recognition, pp. 6309–6318 (2017)
6. Brümmer, N., De Villiers, E.: The BOSARIS toolkit: theory, algorithms and code for surviving the new DCF. arXiv preprint arXiv:1304.2865 (2013)
7. Samhat, A., Altman, Z., Fourestie, B.: Performance analysis of the IEEE 802.11 DCF with imperfect radio conditions. In: 2006 International Conference on Wireless and Mobile Communications, ICWMC 2006, p. 27. IEEE (2006)
8. Salama, A., Zaidi, S.A., McLernon, D., Qazzaz, M.M.: FLCC: efficient distributed federated learning on IoMT over CSMA/CA. arXiv preprint arXiv:2304.13549 (2023)

9. Andersen, P.A.: Deep reinforcement learning using capsules in advanced game environments. arXiv preprint arXiv:1801.09597 (2018)
10. Xu, K., Gerla, M., Bae, S.: How effective is the IEEE 802.11 RTS/CTS handshake in ad hoc networks. In: 2002 Global Telecommunications Conference, GLOBECOM 2002, vol. 1, pp. 72–76. IEEE (2002)

Deep Learning-Based Lung Nodule Segmentation and 3D Reconstruction Algorithm for CT Images

Cheng Xu[1]($^{\boxtimes}$) ⓘ, Shanshan Hua[2] ⓘ, and Meilin Zhong[3] ⓘ

[1] School of Information Engineering, East China Jiaotong University,
Nanchang 330013, China
1079692215@qq.com

[2] Industrial and Commercial Bank of China Bengbu Branch, Bengbu 233000, China

[3] School of Materials Science and Engineering, East China Jiaotong University,
Nanchang 330013, China
zhongmeiling@ecjtu.edu.cn

Abstract. In recent years, deep learning techniques have been widely applied in the field of medical image processing, particularly in the segmentation of pulmonary nodules, which has garnered increasing attention from researchers. The segmentation of pulmonary computed tomography (CT) images is fundamental to the three-dimensional reconstruction of medical images, and the accuracy of image segmentation directly impacts the practical application value of three-dimensional reconstruction in the medical field. However, within the entire CT image of the lungs, pulmonary nodules only occupy a small region, and the complex nature of nodule manifestations and significant size variations present a major challenge for accurate segmentation in modern medicine. To address the issues of precise segmentation of pulmonary nodule CT images and the occurrence of voids in three-dimensional reconstruction, this study focuses on pulmonary CT images and achieves both accurate pulmonary nodule segmentation and three-dimensional reconstruction.

Keywords: Medical CT images · Lung nodule segmentation · 3D reconstruction · Mask R-CNN algorithm · MC algorithm

1 Introduction

World Health Organization (WHO) publicly available data shows that in 2022, there were 2.2 million new cases of lung cancer globally, with a staggering 1.8 million deaths [1]. It's worth noting that the majority of lung cancer patients worldwide are from our country. Over the past decade, both the number of new lung cancer cases and deaths in our country have been steadily increasing. Due

Supported by organizations of the National Natural Science Foundation of China, the Natural Science Foundation of Jiangxi Provincial, and the Major Discipline Academic and Technical Leaders Training Program of Jiangxi Province.

J. Vaidya et al. (Eds.): AIS&P 2023, LNCS 14510, pp. 196–207, 2024.
https://doi.org/10.1007/978-981-99-9788-6_17

to limited medical technology, early detection of lung cancer is crucial for effective treatment. Early-stage lung cancer often presents as pulmonary nodules, making timely detection through medical imaging technology and assessment of their malignant potential critical [2]. Pulmonary nodules typically have diameters ranging from 3 mm to 30 mm and can exhibit various shapes and irregularities in their edges [3].

In recent years, with significant advancements in medical diagnostic and treatment technologies, CT scans have become the most effective method for lung cancer screening [4]. However, the interpretation of CT imaging results still heavily relies on the expertise and clinical experience of doctors. They need to visually assess features such as nodule size and edges, combining these observations with their clinical knowledge to make diagnoses [5]. Even under these conditions, the task remains mentally and physically demanding for doctors, who must review a large volume of CT images within limited timeframes. This can easily lead to false positives (identifying benign tumors as malignant) and false negatives (mistaking malignant tumors for benign ones), resulting in misdiagnosis or delayed treatment. Statistics show that the accuracy of lung cancer CT scans in China is only around 60%–70%, with a misdiagnosis rate of 30%–40% [6].

Medical image segmentation plays a broad and crucial role in medical treatment. It assists doctors in accurately identifying and localizing areas of interest, enabling more targeted treatment plans [7]. However, the variability in size, shape, and distribution of pulmonary nodules, coupled with the subtle differences between nodules and surrounding normal lung tissue in medical images, complicates the task of segmenting pulmonary nodules [8]. Despite these challenges, the use of deep learning algorithms has significantly improved the speed and accuracy of medical image segmentation. Deep learning algorithms, particularly in tasks like pulmonary nodule segmentation, have become mainstream techniques due to the continuous advancements in computer technology [9].

Medical image 3D reconstruction brings tremendous convenience to treatments. By extracting 3D structural information from 2D image sequences using computer technology, doctors can accurately assess lesions before surgery, reducing potential trauma and increasing the success rate of surgeries [10]. Medical image 3D reconstruction is an advancement upon medical image segmentation, aiding doctors in more precise disease diagnosis. Additionally, it holds practical value in areas such as educational training and preoperative planning for reconstructive surgeries. Therefore, leveraging the computational power of computers to achieve pulmonary nodule segmentation and 3D reconstruction holds paramount importance in lung cancer treatment [11]. The main contributions of this article are as follows:

(1) Addressing the complexity of pulmonary nodule features and the high demands for details in lung CT images, this paper proposes an enhanced Mask R-CNN pulmonary nodule segmentation algorithm. By introducing dilated convolutions into the feature pyramid network, image information loss can be reduced, thereby enhancing the accuracy of pulmonary nodule segmentation in CT images.

(2) To tackle the ambiguity present in the traditional MC (Marching Cubes) algorithm during the 3D reconstruction of lung CT images, which results in surface voids in the reconstructed object, this paper presents an improved MC algorithm.

2   Materials and Methods

2.1   Lung Nodule Segmentation Algorithm Based on Enhanced Mask R-CNN

Lung nodules are not only characterized by their complexity but also occupy a relatively small area compared to the entire CT image of the lung. As a result, conventional methods are insufficient for precise segmentation. Based on this premise, this study employs an improved Mask R-CNN algorithm for lung nodule segmentation, demonstrating remarkable segmentation performance.

The backbone network of Mask R-CNN consists of two parts: the first part is ResNet, used to extract high-level semantic features from input images, and the second part is the Feature Pyramid Network (FPN), employed to fuse feature maps from different resolutions to enhance model accuracy and robustness. In Mask R-CNN, FPN receives feature maps from ResNet and generates a series of feature maps with varying resolutions through operations like upsampling and downsampling. These maps are utilized for detecting and segmenting objects of different sizes. However, in a conventional FPN structure (as shown in Fig. 1), performing a 2× upsampling from FPN4 to FPN3 could lead to irreversible loss of image information.

Although each layer's feature maps in FPN undergo feature fusion to reduce information loss, the segmentation of lung nodules remains challenging in the medical field due to their complexity and small size. A notable characteristic of dilated convolutions (also known as atrous convolutions) is the ability to control receptive field size by adjusting convolutional kernel parameters, thus enlarging the effective receptive field. Consequently, the incorporation of dilated convolutions enables feature extraction and representation for objects of different sizes. This paper introduces dilated convolutions into FPN to mitigate image information loss, resulting in improved model performance and accuracy, thereby enhancing lung nodule segmentation.

The advantage of dilated convolutions lies not only in enlarging the receptive field but also in preserving image resolution. This quality makes dilated convolutions an effective choice for segmenting small target objects, concurrently aiding in increasing model precision and stability. By introducing dilated convolutions into FPN, more image information can be integrated into the Feature Pyramid

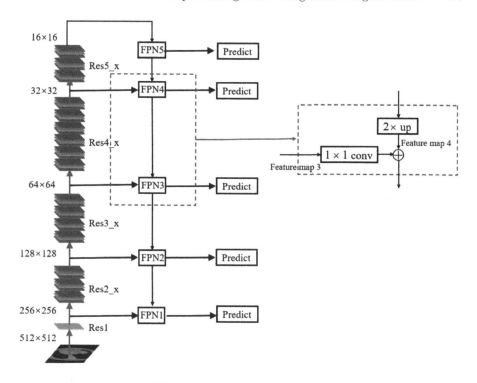

Fig. 1. FPN structure diagram.

Network, significantly enhancing the accuracy of lung nodule segmentation. The modified FPN in this study is referred to as D-FPN, with its structure illustrated in Fig. 2. Taking FPN1 as an example, the feature map of the ResNet's res1 convolutional block is subjected to dilated convolution operations using convolutional kernels with dilation rates of 2, 3, and 4. These operations produce three feature maps, which are then summed to yield D-FPN 1. This version of D-FPN 1 represents fused features of different scales while retaining advanced-level characteristics without altering the size. Subsequently, the obtained D-FPN 1 and feature map1 are added separately to the results obtained from convolutions and 2× upsampling of FPN2. The approach described in this paper replaces the FPN in Mask R-CNN with D-FPN for lung CT image segmentation.

2.2 Three-Dimensional Reconstruction of Medical Images Based on Improved MC Algorithm

In the traditional diagnostic process, doctors rely on their clinical experience and subjective imagination to reconstruct the three-dimensional structure of lesions from a series of medical image data, such as CT and MRI images. In recent years, the rapid development of digital image processing technology has brought new solutions to the field of medicine. Medical image three-dimensional reconstruc-

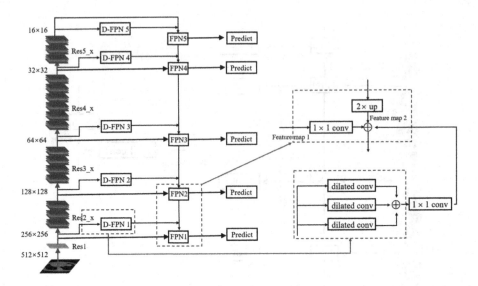

Fig. 2. D-FPN structure diagram.

tion technology can transform medical image data into three-dimensional models with spatial information, thereby helping doctors better understand the morphology, location, and extent of lesions. This technology can also aid in planning surgical and treatment strategies based on specific situations, thus enhancing medical effectiveness and reducing surgical risks.

In the previous section, the improved Mask R-CNN algorithm was used to segment lung nodule CT images, providing a crucial foundation for the three-dimensional reconstruction of lung nodules.

In the field of medical image three-dimensional reconstruction, the Marching Cubes (MC) algorithm has been widely applied to the reconstruction of lung nodules. It is an efficient and reliable method. Despite its capability to generate satisfactory three-dimensional models, the MC algorithm still has some shortcomings. Firstly, it exhibits ambiguity during the reconstruction process, leading to holes in the resulting three-dimensional model. Secondly, the MC algorithm involves the computation of a significant number of empty voxels during the voxel traversal process, which wastes considerable computation time. Thirdly, the three-dimensional models generated by the MC algorithm contain excessive triangular facets, which is unfavorable for model rendering and storage.

Addressing the aforementioned drawbacks, we have made a series of improvements to the MC algorithm. First, during the three-dimensional reconstruction of lung nodule CT images, we traverse the voxels and utilize contour information from neighboring voxels to construct the contour of each voxel. Simultaneously, we have redesigned the 15 triangular facet configurations from the original MC algorithm to eliminate the influence of ambiguity on the three-dimensional reconstruction of the model. These enhancements have led to the creation of 21 new triangular facet configurations (as shown in Fig. 3). Furthermore, our improved MC algorithm significantly reduces computational load by excluding process-ing of empty voxels. Lastly, when calculating the intersection points between contour lines and edges, we employ a simpler median method, further simpli-fying the complexity of the calculation process. These collective improvements result in the superior performance of our MC algorithm in the three-dimensional reconstruction of lung nodule CT images.

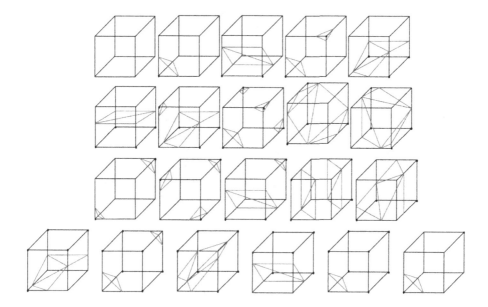

Fig. 3. 21 improved topology models.

3 Experiments and Results

To validate the effectiveness of the proposed lung nodule segmentation method, this study is based on the LIDC-IDRI public dataset. The improved Mask R-CNN algorithm is compared with segmentation models from three reference papers. Among them, reference [12] introduced a novel network model called 3D-Res2UNet, which employs symmetrically layered connectivity to enhance fea-ture extraction capability across multiple scales. Reference [13] chose the Mask R-CNN network for lung nodule segmentation and employed a patch-based app-roach for training the neural network, resulting in more efficient and stable train-ing. Reference [14] utilized deep learning-based image segmentation techniques, developing an enhanced U-Net architecture to learn the distinctive color pat-terns of nodules. To assess the effectiveness of the network models, this study employs evaluation metrics including Dice coefficient, IOU, Recall, and Accu-racy, as shown in Formulas 1, 2, 3, and 4. Here, B_p represents the ground truth for lung nodule segmentation, while B_g represents the segmentation results of lung nodules.

$$\text{Accuracy} = \frac{TN + TP}{FP + TN + TP + FN} \tag{1}$$

$$\text{Recall} = \frac{TP}{TP + FN} \tag{2}$$

$$\text{IOU} = \frac{\text{area}\,(B_p \cap B_g)}{\text{area}\,(B_p \cup B_g)} \tag{3}$$

$$\text{Dice} = \frac{2|A \cap B|}{|A| + |B|} \tag{4}$$

Before commencing the experiments, we partitioned the dataset into training, validation, and test sets in a 6:2:2 ratio. Throughout the experiment, the IOU was set at 0.6. This signifies that during the experiment, if the overlap area between a candidate bounding box and the ground truth label exceeded 0.6, the network regarded it as a positive sample; conversely, it was deemed a negative sample. As for the experiment's hyperparameters, the learning rate was set at 0.001, the weight decay parameter at 0.0001, and the model optimization parameter momentum at 0.9.

Figure 4 illustrates the segmentation outcomes of different network models on the LIDC-IDRI dataset. Furthermore, to further validate the Efficiency of our proposed method, we selected four experimental samples for visual representation, as shown in Fig. 5. It is evident from the figures that our enhanced Mask R-CNN algorithm, in terms of Dice coefficient, recall, and accuracy metrics, outperformed the three network models mentioned in references [12–14]. This substantiates that the improved Mask R-CNN demonstrates remarkable performance in lung nodule image segmentation.

Furthermore, to verify the effectiveness of the MC algorithm proposed in this paper for three-dimensional reconstruction of pulmonary nodules and lung parenchyma CT images, we conducted three-dimensional reconstruction experiments using both the traditional MC algorithm and the improved MC algorithm introduced in this paper. It should be noted that the primary advantage of the three-dimensional reconstruction model lies in the ability to observe lesion areas from multiple angles, a capability not present in two-dimensional CT data sequences.

As depicted in Figs. 6 and 7, the comparative images of three-dimensional reconstruction models obtained through the traditional MC algorithm and the improved MC algorithm from various viewpoints are shown. Figure 6 illustrates the results observed from a frontal perspective, while Fig. 7 presents the results from a lateral viewpoint. In these figures, the left side showcases the reconstruction outcomes of lung parenchyma and pulmonary nodules using the traditional MC algorithm, whereas the right side displays the results obtained using the proposed improved MC algorithm. In the reconstruction of lung parenchyma,

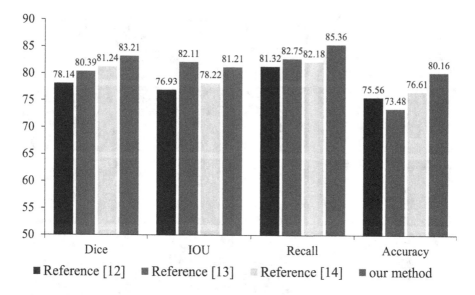

Fig. 4. Performance metrics analysis of four segmentation methods.

it is difficult to discern detailed portions of the pulmonary nodules. However, after segmenting the pulmonary nodules and then conducting reconstruction, the detailed parts of the nodules become visible. By combining the model for lung parenchyma reconstruction, it becomes possible to observe the relative sizes of pulmonary nodules compared to the lung parenchyma, as well as the positions of the nodules. These aspects aid doctors in observing information such as nodule size, characteristics, and density through the three-dimensional reconstruction model, thus facilitating the formulation of a suitable surgical treatment plan.

Employing the improved MC algorithm for three-dimensional reconstruction allows for a clear display of the location of pulmonary nodules within the lungs. The edges of the nodules exhibit well-defined structures and distinct light and shadow effects. This indicates that the improved MC algorithm possesses the advantage of high-quality imaging during the three-dimensional reconstruction process, thereby validating the effectiveness of redesigning triangular mesh configurations using adjacent voxel iso-surface information to address the issue of voids in the reconstruction results. The traditional MC algorithm involves extensive computation of empty voxels during the voxel traversal process, which leads to significant time wastage. The improved MC algorithm proposed in this paper utilizes specific labeling rules to mark the two-dimensional segmentation mask image, excluding a large number of empty voxels containing only blank information in the pulmonary nodule CT images, thereby accelerating the algorithm's execution speed. Furthermore, this paper employs median interpolation instead of linear interpolation to calculate the intersection points between iso-surfaces and voxel edges. This change not only minimally affects reconstruction accuracy

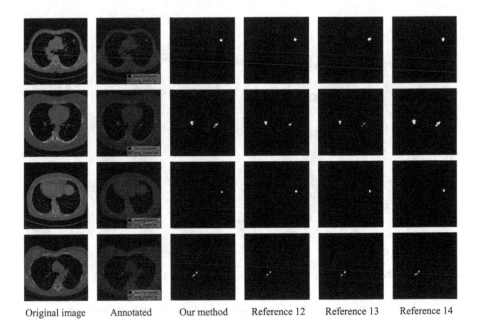

| Original image | Annotated | Our method | Reference 12 | Reference 13 | Reference 14 |

Fig. 5. Display of the segmentation results for four methods.

but also simplifies the complexity of algorithm execution and reduces reconstruction time.

To mitigate the potential stochasticity of experimental results, the present study employed a second set of data for further reconstruction, yielding the reconstructed results shown in Fig. 8. Upon an overall assessment of the reconstruction quality, it is evident that the proposed enhanced MC algorithm, as compared to the conventional MC algorithm, results in a smoother surface recon-

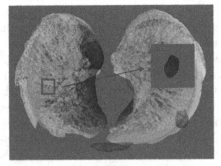 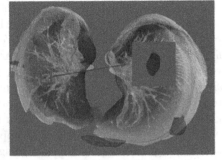

Traditional MC algorithm   Improved MC algorithm

Fig. 6. Comparison of reconstruction results from frontal view.

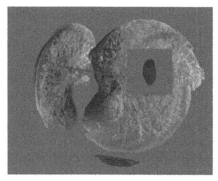
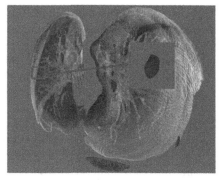

Traditional MC algorithm   Improved MC algorithm

Fig. 7. Side view reconstruction results comparison chart.

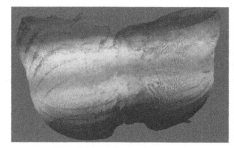
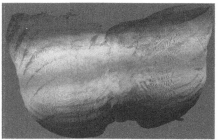

Traditional MC algorithm   Improved MC algorithm

Fig. 8. Comparison of lung 3D reconstruction results in the second dataset.

struction of the three-dimensional model. Therefore, it can be concluded that the proposed improvement method effectively mitigates the impact of voids, leading to more realistic reconstructed model outcomes.

4 Discussion

Early diagnosis and detection of lung cancer are of paramount importance. This article discusses advancements in medical image segmentation and three-dimensional reconstruction methods in relevant fields. However, there are still many areas that require further in-depth research. We outline the following three tasks that need to be pursued:

(1) This paper primarily investigates the application of the Mask R-CNN algorithm in lung nodule segmentation. Building upon this foundation, it is possible to further determine whether a nodule is benign or malignant. This can enhance lung cancer detection, assisting in the early discovery of lung cancer in more patients when feasible.

(2) This paper focuses on enhancing the MC algorithm for three-dimensional reconstruction of medical images. Nevertheless, there is currently no standardized evaluation metric for assessing the quality of three-dimensional reconstruction models. Existing evaluations often rely on visual assessment by the human eye, making them susceptible to subjective impressions. Thus, further research into quantitatively evaluating three-dimensional reconstruction models is imperative.

(3) In light of the current research on three-dimensional reconstruction in this paper, it is feasible to develop a three-dimensional visualization system based on the reconstruction. This system would enable continuous rotation, scaling, and other operations on the reconstructed objects, thereby obtaining more detailed and comprehensive information.

5 Conclusion

This study aims to enhance the segmentation and three-dimensional reconstruction of lung CT images using deep learning techniques. Firstly, the lung CT sequence images are accurately segmented, particularly focusing on the small area of lung nodules, through an improved Mask R-CNN algorithm. To enhance the segmentation performance, we introduce dilated convolutions in the feature pyramid network and validate the superiority of our approach through comparative experiments with other advanced models. Secondly, we enhance the MC algorithm by redesigning the triangle mesh configuration and optimizing the computational process. This successfully addresses the issues of traditional MC algorithms in 3D reconstruction, resulting in a more accurate and efficient reconstructed model. These improvements provide a more intuitive and effective depiction of lung nodule information for clinical diagnosis and offer valuable guidance for further research and applications in the field of medical image processing.

Acknowledgment. This work was supported in part by the National Natural Science Foundation of China under Grant 92159102, the Natural Science Foundation of Jiangxi Provincial under Grant 20232ACB205001 and 20212BAB204005, and the Major Discipline Academic and Technical Leaders Training Program of Jiangxi Province under Grant 20232BCJ22025.

References

1. Siegel, R.L., Miller, K.D., Wagle, N.S., Jemal, A.: Cancer statistics, 2023. CA Cancer J. Clin. **73**(1), 17–48 (2023)
2. Caviezel, C., Kostopanagiotou, K., Puippe, G.D., Werner, R.S., Opitz, I.: 'One-stop-shop' diagnosis and stage-adapted surgical therapy for small nodules of early stage lung cancer in a hybrid operating room. Br. J. Surg. **109**(Suppl. 3), znac185.003 (2022)
3. Davies, B., Ghosh, S., Hopkinson D., Vaughan, R., Wing, C.: Institutional report - thoracic general solitary pulmonary nodules: pathological outcome of 150 consecutively resected lesions (2022)

4. Sathya Udayakumar, A., Keerthi Praveen, B.: Advancements in industrial wastewater treatment by integrated membrane technologies. In: Integrated Environmental Technologies for Wastewater Treatment and Sustainable Development, pp. 369–382 (2022)
5. Hazewinkel, A.D., Bowden, J., Wade, K.H., Palmer, T., Wiles, N.J., Tilling, K.: Sensitivity to missing not at random dropout in clinical trials: use and interpretation of the trimmed means estimator. Stat. Med. **41**(8), 1462–1481 (2022)
6. Mazzone, P.J., Lam, L.: Evaluating the patient with a pulmonary nodule: a review. JAMA **327**(3), 264–273 (2022)
7. Sunil, R., Joies, K.M., Cherungottil, A., Bharath, T.U., Renjith, S.: BitMedi: an application to store medical records efficiently and securely. In: Suma, V., Fernando, X., Du, K.-L., Wang, H. (eds.) Evolutionary Computing and Mobile Sustainable Networks. LNDECT, vol. 116, pp. 93–105. Springer, Singapore (2022). https://doi.org/10.1007/978-981-16-9605-3_7
8. Ale, G.B., Simpson, R., Mertens, C., Santillana, E., Hill, B., Carlo, W., Batra, H.: Shape-sensing robotic bronchoscopy in the diagnosis of pulmonary lesions in children. Authorea Preprints (2022)
9. Yu, L., Wen-Bin, L., Yun-Feng, K., Jian-Feng, D., Meng-Han, W., Yong-Xiang, C.: Identification method of illegal electronic devices based on deep learning and application analysis. Communications Technology (2019)
10. Liu, F., Que, D.: Realization of solid plane cutting system and reconstruction of medical 3D image based on MC algorithm. In: 2nd International Conference on Computer Engineering, Information Science & Application Technology, ICCIA 2017 (2017)
11. Hu, S., Liao, Z., Xia, Y.: Domain specific convolution and high frequency reconstruction based unsupervised domain adaptation for medical image segmentation (2022)
12. Xiao, Z., Liu, B., Geng, L., Zhang, F., Liu, Y.: Segmentation of lung nodules using improved 3D-UNet neural network. Symmetry **12**(11), 1787 (2020)
13. Cai, L., Long, T., Dai, Y., Huang, Y.: Mask R-CNN-based detection and segmentation for pulmonary nodule 3D visualization diagnosis. IEEE Access **8**, 44400–44409 (2020)
14. Du, G., Cao, X., Liang, J., Chen, X., Zhan, Y.: Medical image segmentation based on U-Net: a review. J. Imaging Sci. Technol. (2020)

GridFormer: Grid Foreign Object Detection also Requires Transformer

Dengquan Wu[1]([envelope]) [iD], Hanxin Zheng[1], Zixi Li[1], Xin Xie[1], Tijian Cai[1], and Fengyang Shang[2]

[1] East China Jiaotong University, Nanchang 330000, China
2310555878@qq.com
[2] Columbia University, New York, USA

Abstract. Under the rapid development of China's energy industry, the invasion of foreign objects poses a considerable challenge to the operation and maintenance of power transmission channels. Due to the particularity of the power system, there needs to be more open-source datasets for power transmission foreign object detection, which limits the development of this field and needs to be addressed urgently. Additionally, existing object detection models are often too complex to meet the real-time inference requirements of drones and other terminal devices. To solve these problems, this paper proposes a lightweight object detection model named GridFormer based on a hybrid feature extraction network. This model combines the advantages of convolutional neural networks (CNNs) and transformers, aiming to improve object detection accuracy and real-time performance. Experimental results demonstrate that the proposed model achieves an mAP value 96.78 on a power transmission foreign object dataset. On an NVIDIA GPU 3080, the inference speed can reach 68.7 FPS with only minor loss compared to GhostNet. The model achieves an mAP value of 79.04 on the Pascal VOC dataset, further validating its effectiveness. Compared to GhostNet, the proposed model exhibits superior performance in terms of object detection performance. By addressing the issues above, the GridFormer model is expected to support the development of China's energy industry, improve the efficiency of power transmission operation and maintenance, and promote the development of the object detection field.

Keywords: foreign object detection · Transformer · lightweighting · hybrid features · Grid system

1 Introduction

Electricity, as a crucial essential industry, significantly impacts people's lives and property safety, the national economy, and overall economic development

Supported by the National Natural Science Foundation of China, under Grant No. 62162026, and the Science and Technology Project supported by the Education Department of Jiangxi Province, under Grant No. GJJ210611 and the Science and Technology Key Research and Development Program of Jiangxi Province, under Grant No. 20203BBE53029.

[1]. However, due to long-term exposure to natural elements such as rain, snow, and foreign substances, power transmission lines are prone to issues such as foreign object attachments [2]. If these issues are not addressed promptly, they can compromise the stability of the power grid. Current solutions often involve computer vision techniques to detect foreign objects on power transmission lines. For example, Zhang proposed a method named RCNN4SPTL, which replaces the feature extraction network of Faster RCNN with a more lightweight SPTL-Net, thereby improving the detection speed [3]. Additionally, Wang et al. proposed a method based on SSD for power transmission line detection, studying the effects of different feature extraction networks and network parameters on the accuracy and speed of object detection [4]. Jiang et al. identified bird nests on power transmission lines, cropping the detection results into sub-images and filtering out those not containing bird nests using an HSV color space model, significantly improving detection accuracy [5]. Qiu focused on birds as an object category and proposed a lightweight YOLOv4-tiny network model for bird detection on power transmission lines, providing a basis for preventing bird-related power grid shutdowns [6]. However, in real-world scenarios, foreign objects that can invade power transmission lines include birds, bird nests, balloons, and kites [7]. The high-resolution images captured by drones and other equipment contrast with the relatively small size of these foreign objects, posing a challenge for foreign object intrusion detection. This article proposes a lightweight hybrid object detection network named GridFormer that combines convolution's inductive bias advantages and transformers' global modeling capabilities [8] to achieve good generalization performance on object detection tasks [9,10]. This method strikes a new tradeoff between computational cost and detection accuracy, providing a new, effective solution for foreign object intrusion detection in power transmission lines.

2   Methods

2.1   Model Design

This paper intends to construct a lightweight object detection model based on the CNN-Transformer network, see Fig. 1. It consists of three parts. The first part is a hybrid feature extraction network, which extracts image semantic information; the second part is a feature fusion part, which fuses different levels of feature maps through up-sampling and down-sampling operations to generate a multi-scale feature pyramid. The last part is the classification and localization part, which introduces the auxiliary head design and dynamic label assignment strategy so that the middle layer of the network learns more information by richer gradient information to help train. This model combines the advantages of convolution and transformer to seek Pareto improvement in computational cost and detection accuracy on the object detection task.

In this paper, we use the self-attention mechanism in the deep part of the network, which avoids the Patch division of the feature map due to its small enough

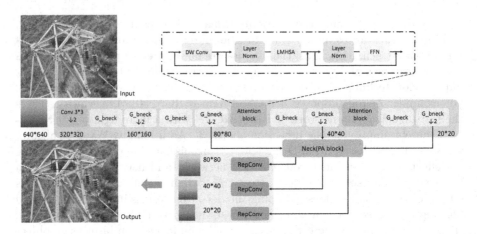

Fig. 1. The overall framework of GridFormer.

size, reduces the effect of positional coding, and ensures the overall consistency of the feature map.

2.2 Attention Block

The structure of the Attention Block is modeled after the design of the Transformer encoder, which is shown in the dashed box in Fig. 1. Absolute position encoding is usually used in ViT, and each patch corresponds to a unique position encoding, so it is not possible to achieve translation invariance in the network. Attention Block uses a 3×3 depth-separated convolution introduces translation invariance of the convolution into the Transformer module, and stabilizes the network training using residual connectivity. Furthermore, the computational procedure is shown in Eq. 1:

$$f(x) = DWConv(x) + x \tag{1}$$

The vital design in the Attention Block is LMHSA. Given an input of size $\mathbb{R}^{n*c}$, the original multi-head attention mechanism first generates the corresponding Query, Key, and Value. Then, by dot-producing the point Query and Key, it produces a weight matrix of size $\mathbb{R}^{n*n}$:

$$Attention(Q, K, V) = softmax(\frac{QK^T}{\sqrt{d}}) * V \tag{2}$$

This process tends to consume a lot of computational resources due to the large size of the input features, making it difficult to train and deploy the network. We use kxk average pooling kernel to downsample the Key and Value branch generation. The calculation process of the lightweight self-attention mechanism is shown in Fig. 2:

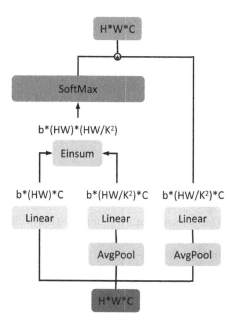

Fig. 2. Calculation process diagram of lightweight self-attention mechanism.

Two relatively small feature maps K' and V' were obtained.

$$K' = AvgPool(K) \in \mathbb{R}^{\frac{n*c}{k^2}} \tag{3}$$

$$V' = AvgPool(V) \in \mathbb{R}^{\frac{n*c}{k^2}} \tag{4}$$

where k is the size and sampling step of the pooling kernel, n is the product of input feature maps H and W, and c is the number of feature map channels. Introducing a pooling layer in the Self-Attention module to downsample the feature map efficiently saves computation and memory. Moreover, LMHSA is composed of multiple Lightweight Self-Attention, see Eq. 5:

$$LMHSA(x) = Concat_{i=[1:h]}[softmax(\frac{Q_iK_i^T}{\sqrt{d}}) * V_i] \in \mathbb{R}^{n*d} \tag{5}$$

The computational complexity of LMHSA is:

$$O(LMHSA) = h * (\frac{n*c^2}{k^2} + \frac{n^2*c}{k^4} + O_\phi(\frac{n^2}{k^2})) \tag{6}$$

where n is the product of the length and width of the output feature map, h is the number of heads of the multi-head self-attention mechanism, k is the kernel size and step size of the pooling kernel, c is the number of channels of the input feature map and the output feature map, and ϕ denotes the softmax activation function.

The computational complexity of the standard self-attention mechanism can be expressed as:

$$O(LMHSA) = h * (n * c^2 + n^2 * c + O_\phi(n^2))$$ (7)

Compared with MHSA, the computational cost of LMHSA is about $\frac{1}{K^2}$ of that of MHSA. The LMHSA in this paper effectively reduces the computational cost, and the optimized matrix operation is more friendly to network model training and inference.

3   Training Methods

In order to improve the accuracy and generalization ability, this paper introduces mixup [13], Mosaic [14] data augmentation, cosine annealing [15] and label smoothing [16] to train the model.

3.1   Data Augmentation

The currently available foreign object dataset has limited capacity, and the size of the sample capacity is crucial for the training effect of the network model. Therefore, we can use data augmentation methods to expand the dataset. Among the commonly used methods for data augmentation include techniques such as spatial transformation and colour transformation. In addition, this paper uses data augmentation methods such as Mosaic and Mixup of image mixing classes. Mosaic data augmentation improves the CutMix method, which aims to enrich the image background while improving the model's detection performance for small objects.

3.2   Cosine Annealing

In order to avoid the model from falling into local optimal solutions, a learning rate adjustment strategy can be used, and one of the standard methods is cosine annealing. The learning rate is gradually reduced through cosine annealing so that the model can better search for the global optimal solution during the training process. This strategy is widely used in deep neural network training and has achieved good results. The learning rate can be reduced by the cosine annealing function, denoted as

$$\eta_t = \eta_{min}^i + \frac{1}{2}(\eta_{max}^i - \eta_{min}^i)[1 + cos(\frac{T_{cur}}{T_i}\pi)]$$ (8)

where η_{max} and η_{min} are the maximum and minimum values of the learning rate, respectively, and T_{cur} and T_i are the current and total number of iterations of an epoch, respectively.

3.3   Label Smoothing

One-hot coded labels in multiclassification problems tend to lead to model overfitting because the model focuses on probability values close to 1. To address this problem, label smoothing can be used to balance the model's predictions and reduce the risk of overfitting. Label smoothing is introduced to smooth the categorical labels, denoted as:

$$y_i' = y_i(1 - \varepsilon) + \frac{\varepsilon}{M} \tag{9}$$

where y_i' is the label after label smoothing, y_i is the one-hot label encoding, M is the number of categories, and ε is the label smoothing hyperparameter.

3.4   Anchor Clustering

The scheme chosen in this paper is an anchor-based object detection model. The traditional anchor selection method is challenging to improve the accuracy, and we first use the K-means clustering algorithm to cluster the manually labelled actual bounding boxes in the training set to obtain the optimal anchor size. Then, we select nine anchor points to predict the bounding box based on the average IoU to improve the detection accuracy. The clustering steps are:

1. Randomly select N boxes as initial anchors;
2. Using the IOU metric, assign each box to the ANCHOR that is closest to it;
3. Calculate the mean value of the width and height of all boxes in each cluster and update the position of the anchor;
4. Repeat steps 2 and 3 until the anchor no longer changes or the maximum number of iterations is reached.

The anchor clustering centers of the transmission line foreign object intrusion dataset in this paper were calculated by K-means as [[38, 49], [81, 55], [61, 88], [78, 153], [110, 120], [144, 98], [155, 182], [200, 257], [317, 380]], as illustrated in Fig. 3. Figure 4 illustrates the Pascal VOC dataset category of anchor clustering.

4   Experiments

4.1   Datasets

Significantly, few open-source datasets are related to the grid intrusion of foreign objects. The foreign object detection dataset used in this paper is mainly from the dataset provided by the 2nd Guangzhou-Pazhou Algorithm Competition-Complex Scene-Based Transmission Corridor Hidden Dangerous object Detection Competition [11], with a total of 800 annotated image data, and the ratio of this paper's training set and test set division is 9:1. among the categories of foreign objects are nest, balloon, kite and trash. This paper also validates the effectiveness of this paper's model on the open-source dataset Pascal VOC [12]. Pascal VOC 2007 has 9,963 images containing 24,640 labelled objects, and Pascal VOC 2012 has 11,540 images containing 27,450 labelled objects, which contain the same 20 object classes for the object detection task.

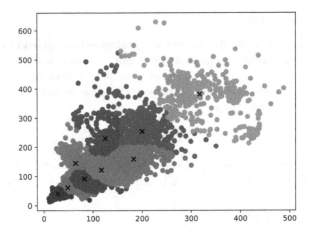

Fig. 3. Anchor clustering of transmission line foreign object detection dataset.

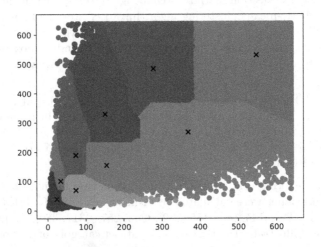

Fig. 4. Anchor clustering of Pascal VOC dataset.

4.2   Experimental Settings

In this paper, the GridFormer model is constructed based on the PyTorch framework, and the neural network parameters are optimized using the Adam optimizer. The initial learning rate is 1e−3, the minimum learning rate is 1e−5, cosine annealing is used to attenuate the learning rate, label smoothing is set to 0.005, the input resolution is 640 * 640, the batch size is set to 8, and the maximum epoch is set to 100. All the training is done using NVIDIA RTX 3080 GPU.

4.3   Evaluation Metrics

In order to reasonably evaluate the performance of the lightweight object detection model, this paper adopts the average value of the APs of each category (mAP) to measure the detection accuracy of the object detection model; the number of floating-point operations, GFLOPs, is used to measure the computational amount of the model, and the number of parameters, Params (M), is used to measure the complexity of the model, which together reflect the computational cost of the model. The number of frames per second, FPS, is used to measure the inference speed of the model. The above four metrics are used to determine the trained model's performance comprehensively.

4.4   Results

On the target dataset of transmission channel hazards, this paper has done relevant experiments and completed the test; the experimental results are shown in Table 1. GridFormer achieved 96.78% mAP with only 25.39M parametric quantities, and the P-R curves of the four categories in the dataset are shown in Fig. 5.

In this paper, we compare the excellent network design of Ghostnet. GridFormer improves the mAP by 4.96% over Ghostnet in terms of accuracy, and the model can reach 68.7 FPS in terms of inference speed, which can satisfy the demand for real-time transmission line foreign object detection.

Table 1. Performance of GridFormer on transmission line foreign object detection dataset

Model	GFLOPs	Params(M)	mAP(%)	FPS
Ghostnet	20.86	26.75	91.82	75.1
GridFormer	21.16	25.39	96.78	68.7

This paper also validates the effectiveness and generalization performance of GridFormer on the Pascal VOC dataset, and the experimental results are shown in Table 2.

Compared with Ghostnet, GridFormer improves the AP on all 15 categories, in which the accuracy of the cow category improves more than 10 AP, and the detection accuracy exceeds 91 AP for both the aeroplane and horse categories. In this paper, we found that the four categories with lower AP of bottle, chair, diningtable, and pottedplant have the superclass of Household, which is due to the cumbersome categories, severe occlusion in household scenarios, and the significant personalized differences of the detected categories, etc. leading to the poor performance in the model inference.

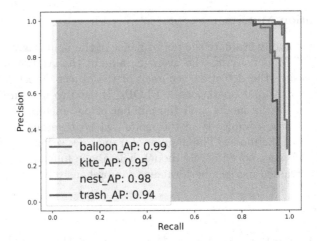

Fig. 5. P-R curves for each category of transmission line foreign object detection dataset.

Table 2. Performance of GridFormer on Pascal VOC dataset

Category	Number of labels	Ghostnet	GridFormer
aeroplane	1336	89.29	**91.17**
bicycle	1259	87.53	**88.64**
bird	1900	82.71	82.58
boat	1294	**71.94**	68.95
bottle	2013	57.62	**59.37**
bus	911	**88.56**	87.43
car	4025	83.29	**86.20**
cat	1768	**85.94**	83.05
chair	3497	61.53	**61.99**
cow	1012	70.77	**81.89**
diningtable	922	61.98	**63.01**
dog	2222	**81.75**	81.15
horse	1287	88.99	**91.14**
motorbike	1248	84.00	**87.95**
person	16051	88.31	**88.86**
pottedplant	1784	53.14	**53.18**
sheep	1129	**84.29**	83.63
sofa	951	71.84	**73.01**
train	1086	82.44	**82.99**
tvmonitor	1283	81.28	**84.56**
mAP	/	77.86	**79.04**

4.5 Ablation Studies

This paper experimentally compares the effect of different training methods on the results. The effects of data augmentation, cosine annealing, and label smoothing on the object detection results are further explored.

Table 3 demonstrates the effect of training methods on model accuracy and F1 scores. In Experiments 2, 3, and 4, Mosaic and Mixup data augmentation techniques were introduced to be able to generate new samples, and the comparison between Experiment 1 and Experiment 2 without the addition of such data augmentation methods showed a growth of 1.4% mAP compared to Experiment 1 without such data augmentation methods. The comparison between Experiment 2 and Experiment 3 shows that labelling improves the model accuracy by 0.6% mAP. In Experiment 4, the label smoothing technique is introduced in this paper, which improves the model accuracy by 1.97% mAP. We believe that this is related to the existence of a more severe data imbalance in the dataset, where there is more data in the category of nest, and the label smoothing technique better handles the scarcity of the samples by adjusting the probability distributions of category labels, thus improving the model's accuracy for all categories. The situation, thus improving the model's ability to detect all categories.

Table 3. Effect of data augmentation, cosine annealing, and label smoothing on the detection.

Exp	method	mAP(%)	F1-Score
1	none	92.80	0.893
2	DA	94.20	0.935
3	DA+CA	94.81	0.933
4	DA+CA+LS	**96.78**	**0.955**

DA: data augmentation; CA: cosine annealing; LS: label smoothing.

5 Conclusion

In grid transmission line systems, foreign object detection is an important protection measure to ensure the regular operation of transmission lines. In this paper, we propose a lightweight object detection model GridFormer based on a CNN-transformer hybrid feature extraction network, which effectively combines the advantages of convolutional induction bias and the transformer's long-term dependence and still can show the excellent performance of the transformer on smaller datasets. The application of GridFormer in the field of transmission line foreign object detection can be better adapted to the scenario of diverse foreign object morphology, smaller objects to be detected and smaller data volume. The experiments show that the model in this paper can find a new tradeoff between inference speed and detection accuracy.

Acknowledgment. This paper is supported by the National Natural Science Foundation of China, under Grant No. 62162026, and the Science and Technology Project supported by the Education Department of Jiangxi Province, under Grant No. GJJ210611 and the Science and Technology Key Research and Development Program of Jiangxi Province, under Grant No. 20203BBE53029.

References

1. Notice of the National Energy Administration on Issuing the "14th Five Year Plan for Electric Power Safety Production". http://zfxxgk.nea.gov.cn/2021-12/08/c_1310442211.htm
2. Ai, Z.: Research on Anomaly Detection Algorithm of Overhead Transmission Lines Based on UAV Aerial Images, Northeast Electric Power University (2021)
3. Zhang, W., Liu, X., Yuan, J.: RCNN-based foreign object detection for securing power transmission lines (RCNN4SPTL). Procedia Comput. Sci. **147**(1), 331–337 (2019)
4. Liang, H., Zuo, C., Wei, W.: Detection and evaluation method of transmission line defects based on deep learning. IEEE Access **8**, 38448–38458 (2020)
5. Hao, J., Wulin, H., Jing, C., Xinyu, L., Xiren, M., Shengbin, Z.: Detection of Bird Nests on Power Line Patrol Using Single Shot Detector, Chinese Automation Congress (CAC), Hangzhou, China, pp. 3409–3414 (2019)
6. Qiu, Z.B., Zhu, X., Liao, C.B.: Detection of bird species related to transmission line faults based on lightweight convolutional neural network. IET Gener. Transm. Distrib. **16**(1), 869–881 (2022)
7. Liu, J.: Research on foreign object detection algorithm and software design of transmission line based on YOLOX, China University of Mining and Technology (2023)
8. Vaswani, A., Shazeer, N., Parmar, N.: Attention is all you need. In: Proceedings of the 31st International Conference on Neural Information Processing Systems, pp. 6000–6010 (2017)
9. Chen, Y.P., Dai, X.Y., Chen, D.D.: Mobile-former: bridging MobileNet and transformer. In: Proceedings of the IEEE/CVF Conference on Computer Vision and Pattern Recognition (CVPR), pp. 5270–5279 (2022)
10. Gulati, A., Qin, J., Chiu, C.-C.: Conformer: convolution-augmented transformer for speech recognition. arXiv:1706.03762 (2021)
11. Transmission Channel Hidden Dangerous Object Detection Algorithm Based on Complex Scenarios. https://aistudio.baidu.com/competition/detail/952/0/introduction
12. Everingham, M., Gool, L.V., Williams, C.K.I., Winn, J., Zisserman, A.: The pascal visual object classes (VOC) challenge. Int. J. Comput. Vis. **88**(2), 303–338 (2010)
13. Zhang, H., Cisse, M., Dauphin, Y.N., Lopez-Paz, D.: mixup: beyond empirical risk minimization. In: ICLR 2018 Conference Blind Submission (2018)
14. Bochkovskiy, A., Wang, C.-Y., Liao, H.: YOLOv4: optimal speed and accuracy of object detection. arXiv:2004.10934 (2020)
15. Xu, G., Cao, H., Dong, Y., Yue, C., Zou, Y.: Stochastic gradient descent with step cosine warm restarts for pathological lymph node image classification via PET/CT images. In: 2020 IEEE 5th International Conference on Signal and Image Processing (ICSIP), Nanjing, China, pp. 490–493 (2020)
16. Wang, C.-Y., Bochkovskiy, A., Liao, H.-Y.M.: YOLOv7: trainable bag-of-freebies sets new state-of-the-art for real-time object detectors. arXiv:2207.02696 (2022)

An Anomaly Detection and Localization Method Based on Feature Fusion and Attention

Zixi Li[✉][ID], Xin Xie[ID], Dengquan Wu[ID], Shenping Xiong[ID], and Tijian Cai[ID]

East China Jiaotong University, Nanchang 330013, China
`zxilee@163.com`

Abstract. In the context of anomaly detection and localization, the small proportion of anomaly data and its unknown nature increase the difficulty of the model in learning the anomaly information. At the same time, the current mainstream detection methods all need help with the problems of limited feature learning ability and weak model generalization ability. Therefore, this paper proposes a multi-module combination of anomaly detection and localization method, which enhances the model's differential learning of normal and abnormal through anomaly simulation strategy and memory module, and fuses multi-scale semantic information through multi-scale feature fusion to improve the accuracy of the judgment of the model on the different scales of abnormality. In addition, the low-cost attention modules acquire global information from multiple perspectives and augment essential features to enhance the feature learning capability. The superiority of the method in this paper is demonstrated in the MVTec anomaly detection dataset.

Keywords: anomaly detection · anomaly localization · feature fusion · attention mechanism

1 Introduction

Anomaly, also known as outliers, mainly refers to the parts that have deviated from the whole, which can be manifested as point anomaly, contextual anomaly, and cluster anomaly [1]. The anomaly detection task is to discover parts that are different from most of the data through various learning methods [2–4]. In computer vision, anomaly detection is primarily performed at the image level, while anomaly localization involves identifying and determining the specific location of anomalies at the pixel level. However, accurately and efficiently accomplishing these anomaly detection tasks poses significant challenges. This is mainly due to the low occurrence probability and high complexity of anomaly data, as well as the limited learning capability of existing detection methods.

With the rapid development of deep learning technology, anomaly detection methods based on deep learning technology have gradually become mainstream methods-the dominant methods based on three categories: classification, reconstruction, and distribution [5]. Classification-based methods classify data mainly

J. Vaidya et al. (Eds.): AIS&P 2023, LNCS 14510, pp. 219–228, 2024.
https://doi.org/10.1007/978-981-99-9788-6_19

using techniques such as clustering and convolutional neural networks, and the deep learning model will mark normal data as one class and abnormal data as another. The deep support vector data description (SVDD) [6] method maps normal data features into a hypersphere and determines the data out of the hypersphere as anomalous. Reconstruction-based methods focus on fully learning the features of normal data through models such as autoencoders to reconstruct normal data well, whereas reconstructing of anomalous data produces large errors. Shi et al. [7] proposed an effective unsupervised anomaly segmentation method for detecting and segmenting anomalies in small and restricted regions of an image, and they designed a deep and efficient convolutional autoencoder for detecting anomalous regions in an image through fast feature reconstruction. Distribution-based methods aim to build a distribution of data through models such as the gaussian mixture model, generative adversarial network (GAN) [8], and normalizing flow (NF) [9] to portray the distribution of the normal data that the normal data is within the distribution, and the anomalous data outside of the distribution. The conditional normalizing flows model [10] consists of a discriminative pretraining encoder and a multiscale generative encoder decoder, which employs a conditional normalized flow framework for anomalies for localization detection.

Although the above methods can solve the problem of anomaly detection and localization in various industries, they still have problems such as sensitivity to noisy data, omission of detection due to normal reconstruction of anomaly data, and weak generalization ability to unknown anomalies. Therefore, this paper proposes a method that combines memory modules, feature fusion, and attention modules for anomaly detection and localization.

2 Proposed Approach

2.1 Network Architecture

The paper presents the overall framework of the proposed model, which is depicted in Fig. 1. The model accepts inputs from normal and anomaly samples generated through anomaly simulation. Anomaly simulation is performed to obtain anomaly samples in both texture and structure through Perlin noise, texture image dataset (DTD) [11], and various transformation operations. In addition, N normal samples are extracted by the pre-trained encoder to obtain high-level features with dimensions of $64 \times 64 \times 64$, $128 \times 32 \times 32$ and $256 \times 16 \times 16$, respectively, which together form the memory features, and N is equal to 30 in this paper. In the training and inference stage, the input and memory features are subjected to L_2 distance calculation to obtain the difference information DI, which is calculated as shown in (1). The best difference information DI^* is obtained by taking the elements in each DI. The minimum is the criterion, as shown in (2). Subsequently, the cascade operation is performed in the channel dimension by combining the high-level features of the input image with the memory sample features associated with the most informative differences. This process results in the generation of cascading features across all three dimensions.

$$DI = \bigcup_{i=1}^{N} \|MF_i - IF\|_2 \tag{1}$$

$$DI^* = \arg\min_{DI_i \in DI} \sum_{x \in DI_i} x, \quad i \in [1, N] \tag{2}$$

The initial layer feature map and the final layer features extracted by the encoder are employed in the decoder for reuse. These features are then combined with the features obtained from multi-scale feature fusion and squeeze-and-excitation attention during the upsampling process. This integration facilitates effective anomaly segmentation and localization.

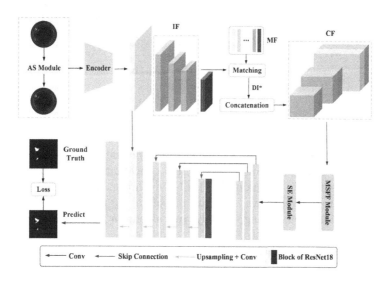

Fig. 1. Model architecture

2.2 Attention Mechanism

Coordinate attention (CA) [12] is a mechanism that preserves precise location information while capturing long-range dependencies by aggregating features from both channel and spatial directions. The resultant weights are subsequently utilized to enhance the feature representation capability of the model by applying them to the input. The process of coordinate attention can be decomposed into two distinct steps: coordinate information embedding and coordinate attention generation. Figure 2 provides an illustrative depiction of the structure of this mechanism.

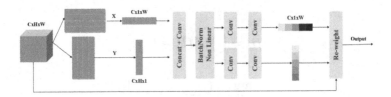

Fig. 2. Coordinate attention

The coordinate information embedding involves utilizing pooling kernels with dimensions $(1, W)$ and $(H, 1)$. These kernels are used to obtain two direction-aware feature maps: one containing global receptive fields and the other preserving precise location information. For an input image X, the output of the cth channel at height h is represented by (3), and the output of the cth channel at width w is represented by (4).

$$Z_c^h(h) = \frac{1}{W} \sum_{0 \leqslant i \leqslant W} X_c(h,i) \tag{3}$$

$$Z_c^w w = \frac{1}{H} \sum_{0 \leqslant j \leqslant H} X_c(j,w) \tag{4}$$

During the coordinate attention generation, the feature maps obtained from the information embedding operation are first concatenated. Subsequently, these concatenated feature maps undergo a 1×1 convolution operation. The resulting feature maps are split into two separate tensors along the spatial dimension to facilitate individual processing. Finally, two attention weights are derived and applied to the input image.

The squeeze-and-excitation (SE) attention module [13] is designed to dynamically learn the correlation between channels to enhance essential features while suppressing less important ones. The SE attention module consists of two primary components: squeeze and excitation. The structure of this module is visually presented in Fig. 3.

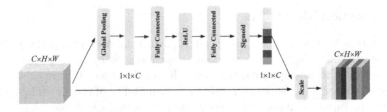

Fig. 3. SE attention

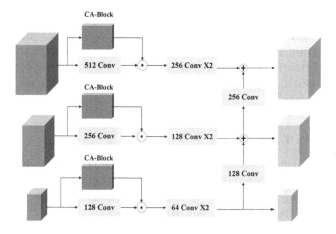

Fig. 4. Multi-scale feature fusion

The squeeze operation involves compressing the information of each channel into a scalar z_c through global average pooling. This results in obtaining a feature vector of dimension $1 \times 1 \times C$, where C represents the number of channels in the input feature map. The calculation is represented by (5).

$$z_c = \frac{1}{W \times H} \sum_{i=1}^{W} \sum_{j=1}^{H} x_c(i, j) \tag{5}$$

where H and W represent the height and width of the input feature map, respectively. x_c refers to the feature matrix of the cth channel. The excitation operation comprises fully connected layers and activation functions. The fully connected layer is responsible for modulating the feature transformation of the vector, while the activation function restricts the output weight s within the range $[0, 1]$. The calculation is demonstrated in (6).

$$s = \sigma\left(W_2 \delta\left(W_1 z\right)\right) \tag{6}$$

where W_1 and W_2 are the weights of the fully connected layers, δ denotes the ReLU activation function, while σ represents the sigmoid activation function. The scale operation is performed as a final step, where the input feature maps are weighted based on the computed channel weights.

2.3   Multi-scale Feature Fusion

To address challenges such as varying anomaly scales, this paper extensively leverages the multi-dimensional feature information generated after the cascade operation, as depicted in Fig. 4.

In the proposed approach, the feature information from different dimensions is initially weighted using the CA mechanism. Additionally, the number of channels is adjusted through convolutional operations, while the image dimensions at different resolutions are modified using upsampling operations. Subsequently, an element-wise summation operation is performed to achieve multi-scale fusion. The fused feature information is then individually passed through the corresponding SE attention module to obtain channel attention weights. This allows the model to emphasize important information while reducing reliance on irrelevant or noisy data.

2.4 Objective Function

To ensure the similarity between the model's predictions (P) and the ground truth (G), this paper minimizes both the L_1 loss and focal loss. The loss are calculated as shown in (7) and (8). These loss functions help optimize the model's output to match the desired ground truth closely.

$$L_1 = \|G - P\|_1 \tag{7}$$

$$L_f = -\alpha_t (1 - p_t)^\gamma \log(p_t) \tag{8}$$

During the training process, when the true label value G of a pixel is 1 (indicating an anomaly sample), p_t is set to the predicted probability p of that pixel's category. Conversely, when the true label value is 0 (representing a normal sample), p_t is defined as $1 - p$. Hyperparameters α_t and γ are introduced to control the weighting degree. These constraints are then integrated into an objective function, as illustrated in (9). The paper aims to minimize this objective function to optimize the model during training. The balance hyperparameters, W_1 and W_f, are set to 0.6 and 0.4, respectively.

$$L_{all} = W_1 L_1 + W_f L_f \tag{9}$$

3   Experiments

In order to evaluate the anomaly detection model effectively, this paper performs experiments on the MVTec AD [14] dataset. MVTec AD is a commonly utilized dataset for surface anomaly detection tasks. To accurately assess the model's effectiveness in anomaly detection and localization, image-level AUROC and pixel-level AUROC are employed as evaluation metrics. Image-level AUROC measures the model's ability to detect anomalies, while pixel-level AUROC evaluates the model's performance in localizing anomalies. These metrics are widely used in anomaly detection tasks for evaluating the efficacy of both anomaly detection and localization capabilities.

3.1 Anomaly Detection

Our model was benchmarked against existing anomaly detection methods in the anomaly detection task. The experimental comparison results on the MVTec AD dataset are presented in Table 1. Our method achieved an average anomaly detection accuracy of 98.39% across 15 categories. Our model outperformed other methods in 9 of the 15 categories, demonstrating its superior performance.

Table 1. The results of anomaly detection with different methods (Image-AUROC (%))

Class	DevNet [15]	UniAD [16]	FYD [17]	DRAEM [18]	ours
Bottle	99.3	**100**	**100**	99.2	**100**
Cable	89.2	97.6	94.3	91.8	**97.75**
Capsule	86.5	85.3	93.2	**98.5**	98.21
Carpet	86.7	**99.9**	98.3	97.0	93.18
Grid	96.7	98.5	97.4	99.9	**100**
Hazelnut	**100**	99.9	99.8	**100**	99.85
Leather	99.9	**100**	**100**	**100**	**100**
Metal_nut	99.1	99.0	99.9	98.7	**100**
Pill	86.6	88.3	94.9	**98.9**	89.57
Screw	97.0	91.9	89.7	93.9	**99.97**
Tile	98.7	99.0	95.4	99.6	**100**
Toothbrush	86.0	95.0	99.9	**100**	98.88
Transistor	92.4	**100**	99.7	93.1	98.37
Wood	99.9	97.9	99.8	99.1	**100**
Zipper	99.0	96.7	97.0	**100**	**100**
Avg	94.5	96.6	97.3	98.0	**98.39**

Bold fonts denote the best performance.

3.2 Anomaly Localization

The experimental results of anomaly localization for our model on the MVTec dataset are displayed in Table 2. Our model attains an average pixel-level AUROC of 91.92% across 15 categories. This demonstrates the effectiveness of our model in accurately localizing anomalies within the images.

Table 2. The results of anomaly localization with different methods (Pixel-AUROC (%))

Class	KDAD [19]	DevNet [15]	ours
Bottle	96.3	95.1	**97.07**
Cable	82.4	**92.0**	90.29
Capsule	**95.9**	93.8	94.75
Carpet	95.6	**96.3**	95.33
Grid	91.8	93.5	**93.59**
Hazelnut	94.6	**95.9**	91.31
Leather	98.1	**99.0**	98.80
Metal_nut	86.4	**87.6**	78.42
Pill	89.6	85.9	**89.71**
Screw	**96.0**	89.7	91.78
Tile	82.8	95.0	**96.27**
Toothbrush	**96.1**	81.9	95.26
Transistor	76.5	**83.9**	81
Wood	84.8	90.0	**93.36**
Zipper	93.9	**97.3**	91.87
Avg	90.7	91.8	**91.92**

Bold fonts denote the best performance.

4   Conclusion

Aiming at the problems of low detection accuracy and weak feature learning ability of existing methods, this paper proposes an image anomaly detection and localization method that combines memory module, attention module, and feature fusion. The anomaly simulation strategy and memory module in this method help the model to learn the difference between normal and anomalous features. The attention mechanism and feature fusion module help the model to comprehensively learn the anomalies of different scales, as well as the importance of the anomalous features so that the model can efficiently detect the anomalies of different sizes and improve the generalization ability of the model. The method in this paper performs well in the MVTec dataset. However, the accuracy of anomaly localization needs to be further improved, which will be the main direction of future research.

Acknowledgment. This paper is supported by the National Natural Science Foundation of China, under Grant No. 62162026, and the Science and Technology Project supported by the Education Department of Jiangxi Province, under Grant No. GJJ210611 and the Science and Technology Key Research and Development Program of Jiangxi Province, under Grant No. 20203BBE53029.

References

1. Pimentel, M.A., Clifton, D.A., Clifton, L., Tarassenko, L.: A review of novelty detection. Sig. Process. **99**(6), 215–249 (2014)
2. Yang, F., Peng, Y., Li, Y.: Research on insulator self-explosion detection with small sample based on deep learning. J. East China Jiaotong Univ. **2**, 110–117 (2022)
3. Walluscheck, S., Canalini, L., Klein, J., Heldmann, S.: Unsupervised learning of healthy anatomy for anomaly detection in brain CT scans. In: Society of Photo-Optical Instrumentation Engineers (SPIE) Conference Series, vol. 12465, p. 1246504 (2023)
4. Tong, H.Z.: Research on multiple classification detection for network traffic anomaly based on deep learning. In: 2022 6th International Symposium on Computer Science and Intelligent Control (ISCSIC), pp. 12–16, November 2022
5. Tao, X., Gong, X., Zhang, X.Y., Yan, S., Adak, C.: Deep learning for unsupervised anomaly localization in industrial images: a survey. IEEE Trans. Instrum. Measur. **71**, 1–21 (2022)
6. Ruff, L., Vandermeulen, R.A., Görnitz, N., Deecke, L., Kloft, M.: Deep one-class classification. In: International Conference on Machine Learning, pp. 4393–4402, July 2018
7. Shi, Y., Yang, J., Qi, Z.: Unsupervised anomaly segmentation via deep feature reconstruction. Neurocomputing **424**, 9–22 (2021)
8. Goodfellow, I.J., et al.: Generative adversarial nets. In: Advances in Neural Information Processing Systems, vol. 27 (2014)
9. Dinh, L., Sohl-Dickstein, J.N., Bengio, S.: Density estimation using Real NVP (2016). https://arxiv.org/abs/1605.08803
10. Gudovskiy, D., Ishizaka, S., Kozuka, K.: CFLOW-AD: real-time unsupervised anomaly detection with localization via conditional normalizing flows. In: Proceedings of the IEEE/CVF Winter Conference on Applications of Computer Vision, pp. 98–107 (2022)
11. Cimpoi, M., Maji, S., Kokkinos, I., Mohamed, S., Vedaldi, A.: Describing textures in the wild. In: Proceedings of the IEEE Conference on Computer Vision and Pattern Recognition, pp. 3606–3613 (2014)
12. Hou, Q.B., Zhou, D.Q., Feng, J.S.: Coordinate attention for efficient mobile network design. In: Proceedings of the IEEE/CVF Conference on Computer Vision and Pattern Recognition, pp. 13708–13717 (2021)
13. Hu, J., Shen, L., Albanie, S., Sun, G., Wu, E.: Squeeze-and-excitation networks. IEEE Trans. Pattern Anal. Mach. Intell. **42**, 2011–2023 (2017)
14. Bergmann, P., Fauser, M., Sattlegger, D., Steger, C.: MVTec AD—a comprehensive real-world dataset for unsupervised anomaly detection. In: Proceedings of the IEEE/CVF Conference on Computer Vision and Pattern Recognition, pp. 9584–9592 (2019)
15. Pang, G., Ding, C., Shen, C., Hengel, A.V.: Explainable deep few-shot anomaly detection with deviation networks (2021). https://arxiv.org/abs/2108.00462
16. You, Z.Y., Cui, L., Shen, Y., Yang, K., Lu, X., Zheng, Y.: A unified model for multiclass anomaly detection. In: Advances in Neural Information Processing Systems, vol. 35, pp. 4571–4584 (2022)
17. Zheng, Y., Wang, X., Deng, R., Bao, T., Zhao, R., Wu, L.: Focus your distribution: coarse-to-fine non-contrastive learning for anomaly detection and localization. In: 2022 IEEE International Conference on Multimedia and Expo (ICME), pp. 1–6 (2022)

18. Zavrtanik, V., Kristan, M., Skočaj, D.: DRAEM – a discriminatively trained reconstruction embedding for surface anomaly detection. In: Proceedings of the IEEE/CVF International Conference on Computer Vision, pp. 8310–8319 (2021)
19. Salehi, M., Sadjadi, N., Baselizadeh, S., Rohban, M.H., Rabiee, H.R.: Multiresolution knowledge distillation for anomaly detection. In: Proceedings of the IEEE/CVF Conference on Computer Vision and Pattern Recognition, pp. 14897–14907 (2021)

Ensemble of Deep Convolutional Network for Citrus Disease Classification Using Leaf Images

Bo Li[1], Jinhong Tang[2(✉)], and Nengke Xie[1]

[1] School of Electronics Information Engineering, Shanghai Dianji University,
Shanghai, China
libo@sdju.edu.cn
[2] School of Electronics and Information Engineering,
Jiangxi Industry Polytechnic College, Nanchang, China
jhtang1997@163.com

Abstract. Crop disease is a major threat to agricultural production. Reduced yield due to crop diseases can lead to immeasurable economic losses. Therefore, the detection and classification of crop diseases are of great significance. Based on AlexNet, VggNet, ResNet, and DenseNet, this paper presents an Ensemble Network improve the recognition accuracy. In this study, firstly training four CNN models by randomly sampling the dataset. Before that, these CNN were improved by batch normalization and global average pooling to accelerate convergence, designed a dynamic concatenated ReLU, an improved activation function on ReLU, to improved detection performance. We use focal loss to solve data imbalance. Then weighted voting is used to fuse the four CNN models. Finally, we verify that the EnsembleNet can effectively improve the recognition performance compared with a single network. Verified improvement in recognition performance on our datasets, obtained from PlantVillag.org, containing 4577 citrus leaf images of three categories. The maximum test accuracy in identifying citrus leaf diseases was as high as 93.58%. Compared with single network, our EnsembleNet can significantly performance. The experimental results showed that this method can be practically applied to the identification of citrus leaf diseases and provides a basis for the identification of other plant leaf diseases.

Keywords: Convolution Neural Network · Crop Disease · Image Classification · Deep Learning

1 Introduction

Crop diseases are the main factors affecting modern agricultural production. Crop diseases not only lead to crop yield reduction and poor quality, but also have an impact on local finance and agricultural economy of a region [2]. Therefore, many experts have conducted in-depth studies on preventive measures. Because crop diseases show many symptoms, most of them occur on the leaves,

J. Vaidya et al. (Eds.): AIS&P 2023, LNCS 14510, pp. 229–240, 2024.
https://doi.org/10.1007/978-981-99-9788-6_20

judging the occurrence of diseases from the deterioration degree of leaf has always been a research direction.

Classical machine learning requires an additional stage of data preprocessing as feature engineering is very crucial, which requires engineers to understand the features of the data set in depth. Moreover, classical machine learning depends on the data in hand. An imbalanced, small dataset results in poor accuracy, but after reaching certain accuracy, the increase in the dataset does not affect the performance much. Our work faces the same problems as well. CNNs models overcome these problems, especially the datasets composed of images. Introducing deep learning into image classification can be effective. In recent years, researchers have made significant achievements in this field [1,3,6,15,17,18]. Unlike the classic machine learning, the CNN models does not need to manually extract or clean data. It can learn statistical rules from a large number of training samples and automatically extract the features. Despite crop disease detection based on artificial intelligence has vast number of studies in the past few years.

However, most of the research is a single network that identifies crop diseases, In this work. In this study, 4577 images of raw citrus (with disease and healthy) leaves were trained using an ensemble CNN model, including four CNN models (AlexNet [13], VggNet [16], ResNet [8], and DenseNet [10]), Each of them will use random sampling method to select training samples from the dataset, which ensure that the training of CNN models is not exactly the same. Random sampling makes the training of the four CNN models more independent. For each CNN models to be more reliable, we take some strategies for each CNN model. First, Using dynamic concatenated ReLU (CDy-ReLU), an improved activation function on ReLU. Then use focal loss [14] to solve data imbalance and compare the accuracies of several optimization methods. Finally, we combine the four Trained models into a EnsembleNet and compare the performance of the EnsembleNet with four CNN models. The four CNN models are developed into integrated model by weighted voting, and the voting weights are initialized to average value. Notably, the performance of the integrated network on citrus leaf dataset is better than that of four CNN models.

The main contributions of this paper are as follows:

(1) A method based on CNN architecture is proposed to realize Citrus diseases identification. This method exploits Ensemble Learning to improve recognition performance.
(2) This paper presents a dynamic activation function, which is improved on the basis of ReLU.
(3) A newly Citrus diseases dataset is published in this paper. The dataset contains 31938 images.

2   Materials and Methods

2.1   Dataset

The citrus leaf dataset used in this study was obtained from PlantVillage, an open access public resource for agriculture-related content. The dataset includes three types of citrus leaves: citrus healthy, citrus HLB (Citrus yellow shoot) general, and citrus HLB serious. The original dataset contains 4577 images of citrus leaves, divided into three categories, as shown in Table 1 and Fig. 1.

By counting the total number of samples and the distribution of various samples, it is found that the distribution of the sample number is unbalanced. The deep CNN used in this study contains millions of parameters. Uneven or too few samples may lead to over-fitting or lack of robustness of the deep neural network. To solve the data imbalance problem, the next step is to enrich the dataset with data augmentation. When training a network, the more the images of a category, the more chances the network will learn the features of the category, and the classification accuracy will be improved. To enrich the dataset with augmented images, a database of 31938 training images and 6320 validation images is created, as explained in next section.

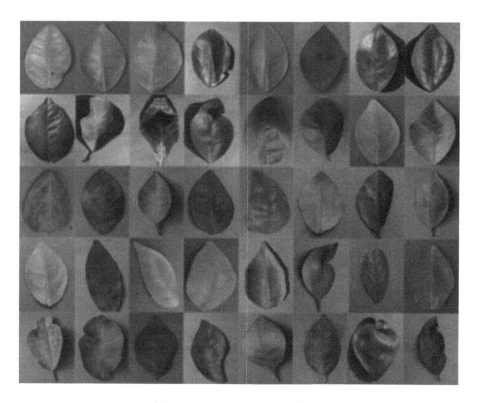

Fig. 1. Raw citrus leaf images

Table 1. Number of citrus leaf images in each category

Class (Scientific name)	Number of images
Citrus healthy	412
Citrus healthy	2097
Citrus HLB serious	2061

2.2 Data Augmentation

In this work, we use AutoAugment [5] for image data augmentation. The main idea of this method is to create a search space for data enhancement and to evaluate the quality of the specific policies directly on the dataset. In this experiment, we have a search space, each policy is composed of many sub-policies, and a sub-strategy is randomly selected for each image in each batch. A substrategy consists of two operations, each being an image processing method, such as translation, rotation, or cutting. Each operation has a type, amplitude, and probability.

To enhance the data of the raw citrus leaf images, we use the transfer learning strategy, initialize the network with the weights in the image network [5], and finally adjust these weights to obtain a combination of 25 sub-policies.

The augmented dataset contains 31938 images; the specific distribution is listed in Table 2. The following experiments were conducted for the augmentation.

Table 2. Augmented citrus leaf dataset

Class (Scientific name)	Number of images
Citrus healthy	10410
Citrus healthy	10854
Citrus HLB serious	10674

2.3 Concatenated Dynamic ReLU

ReLU (Rectified Linear Unit) is one of activation function in artificial neural networks, It is widely used because of simplicity and efficiency. However, there is an obvious drawback to ReLU. During training, some neurons will dead. That they stop outputting anything other than 0. From the mathematical formula of this function, $f(x) = \max(0, x)$. From this formula we can easily know, a negative gradient is zeroed in the ReLU function. This means that the activation function only retains the positive phase and ignore the negative phase information. In order to solve this problem, researchers have proposed a variety of ReLU variants

(LeakReLU, PReLU [7], ELU [4] and SELU [12]) but mostly static. In 2020, Microsoft proposed a Non-static activation function, dynamic ReLU (Dy-ReLU), Its parameters are determined by all the input data. The key principle is that Dy-ReLU encodes the global context information into its function enhancement function, and makes its piecewise linear function adapt to each other. Compared with static ReLU, Dy-ReLU, it only increases little computation but significantly enhances its characterization ability.

This paper proposed a new way to improve the dynamic ReLU, and verify in Sect. 3 that it is effective. Verify that the lower convolution kernel of the network is more inclined to capture the information of positive and negative phase. Therefore, we define and verify a new ReLU, concatenated dynamic ReLU (CDy-ReLU) by combining with dynamic ReLU

$$CDy\text{-}ReLU(x) = [DReLU(x), DReLU(-x)] \tag{1}$$

Dynamic Concatenated ReLU dynamically adjusts its parameters based on the input's sign, making neurons more adaptive to diverse inputs. This improvement significantly enhances feature extraction, especially in the recognition of fine details within plant leaf disease images, ultimately leading to improved detection accuracy.

2.4   Ensemble Network

The Ensemble Network is divided into two stages. As shown in Fig. 2, in stage one, four CNN models Sampling from data sets as training, and the method of sampling is random sampling. Each time we randomly collect a sample and put it into the sampling set, and then put the sample back. The sample may still be collected in the next sampling. After collecting m times, we can finally get the sampling set of m samples. Because it is random sampling, the sampling set of each time is different from the original training set and other sampling sets. We know that the original dataset is imbalanced. We use AutoAugment to enhance the dataset. Through the experiment, we know that the characteristics of each category are still imbalanced. A small number of categories in the original dataset belong to hard sample, so we use focal loss to balance the categories of the dataset. We add a BN layer after the convolution layer to improve the convergence rate and replace the fully connected layer with global average pooling to reduce the number of parameters. The functions of these three layers are as follows: (1) Batch Normalization (BatchNorm) is an essential regularization technique when training deep neural networks. It aids in mitigating the issues of gradient explosion and vanishing gradients, making the network more amenable to training. By normalizing the inputs of each layer within every training batch, BatchNorm helps stabilize the internal distribution of the network, accelerating convergence, and enhancing model generalization. In our study, we incorporated BatchNorm layers into four CNN models to ensure their stability during training, further improving their performance. (2) Global Average Pooling (GAP): Traditional convolutional neural networks often employ fully connected layers in the final layer to produce the ultimate classification results. However, fully connected layers come with a high parameter count, making them prone to over-fitting. In contrast, the Global Average Pooling layer computes the average of feature maps for each channel, generating a global feature vector used for classification. This approach reduces the number of parameters, thereby reducing the risk of overfitting. By implementing Global Average Pooling in the four CNN models, we decreased model complexity and enhanced generalization. (3) Dynamic Cascading ReLU Activation Function: Activation functions play a crucial role in neural networks, determining the level of activation of neurons. In our research, we introduced a dynamic cascading ReLU activation function as an enhancement of the traditional ReLU activation function. The dynamic cascading ReLU adjusts activation function parameters adaptively based on the input's polarity, rendering neurons more responsive to various inputs. This improvement aids in enhancing the network's feature extraction capability, strengthening the recognition performance of plant leaf disease characteristics. This enhancement allows the model to better capture minor features in plant leaf images, thereby improving disease detection accuracy.

In the first stage, four models were trained to predict the diseases of citrus leaf. In the second stage, the task of this work is to use mathematical technology to combine models with voting. This simple and effective approach can be understood as the ensemble method is taking more than one expert opinion and determines its decision according to the majority of experts. The voting technique in this job is not majority voting, and simple majority voting techniques might produce inaccurate results. After all, the prediction of CNN models is based on the probability method.

Recall that each CNN model calculates a probability value for each class in the task, and based on this, assigns a class tag with the highest probability value to the image. When data sets and class are related. The outputs of the classification layer of the model, namely the probability set, $p^i = [p_1^i, ..., p_n^i]$, and not doubt that $\sum_{j=1}^n p_j^i = 1$. Thus, when input a test image to EnsembleNet, the prediction $OP = [P_1, ..., P_n], \sum_{c=0}^n P_c = 1$. The task of our work is transformed into getting probability set OP.

$$OP = [P_1, ..., P_n], and, P_c = \frac{\sum_{i=1}^m p_i^c}{m}, c = 1, ..., n \qquad (2)$$

Here, is the number of CNN models, in this paper. Therefore, the prediction of the test image is:

$$predicition = \max(OP) = \max(P_c), c = 1, ..., n \qquad (3)$$

CNN models in the EnsembletNet can be regarded as experts, and each expert will propose a probability set to describe this possibility. EnsembletNet determines the category of test image by means of weighted voting, with the weight of the output of each CNN.

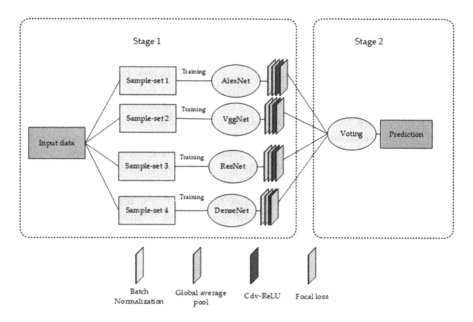

Fig. 2. Raw citrus leaf images

3   Experimental Results

The experiment is divided into three parts: the first part compares the training of four CNN models, including AlexNet, VGG 16, ResNet 50, and DenseNet 201. In the second part, in order to evaluate the performance of the proposed activation function, CDy-ReLU, we compared the traditional ReLU and CDy-ReLU. The same data is executed on the same CNN model, with traditional ReLU and with CDy-ReLU. In the first part of the test set, we tested four CNN models, and after ten tests, calculated their test accuracy. In the third part, we compare the performance of single network and EnsembleNet.

3.1   Influence of Single Network Accuracy and Data Imbalance

In this experiment, we used $256 \times 256 \times 3$ images. To evaluate the difference between the real sample label and the prediction label of the network, we take the loss function and training accuracy as the evaluation criteria for the neural network performance. The lower the loss function, the closer is the predicted value of the network to the real value, and the better is the robustness of the network. Moreover, the higher the training accuracy, the better is the network performance.

In terms of the evaluation criteria of the model, we choose to conduct 10-fold cross validation on the data set to obtain the recognition accuracy. Specifically, the data set is divided into ten parts, and nine of them are tested as training data and one as test data in turn. Each test will get the correct rate. The average accuracy of the results of 10 times is used as an estimate of the accuracy of the algorithm.

In order to compare the accuracy of a single network on the test set, we did some experiments to compare them. Table 3 indicates that, when the other parameters are the same, the models with BN have a better test accuracy than without the BN. The best test accuracy of ResNet 50 model (86.55%) is approximately 2% points more than that of the model without the BN, DenseNet201 (86.02%), improving by nearly 4% points when using the BN. The accuracy of the other two models with BN is also improved to varying degrees, AlexNet (83.28%) and VggNet19 (85.62%). Thus, it is proven that BN can improve the recognition accuracy on the test set.

In addition, during the experiment, we found that even if image enhancement has influence on the data imbalance, the recognition accuracy of each category will still be affected by the data imbalance. In our opinion, image enhancement can't solve the problem that the features of less sample categories are less than those of more sample categories. As shown in Fig. 3(a), The confusion matrix can clearly show that the recognition accuracy of these three categories is not balanced. Each column of the confusion matrix represents the predicted category, and the total number of each column represents the amount of data predicted for that category. Each row represents the real data category, and the total data in each row represents the number of data instances in this category. It can be seen from Fig. 3(a) that the recognition accuracy of healthy citrus leaves is only 81%. Although the recognition accuracy rate of the other two categories reached 96.5% and 96.25% respectively, we can't believe that this model can adapt to the characteristics of citrus disease.

Table 3. Models test set recognition accuracy

Models	with without BN	Number of Layers	Test Accuracy
AlexNet	with BN	8	83.28%
	without BN		82.12%
VggNet19	with BN	19	85.62%
	without BN		83.53%
ResNet50	with BN	50	86.55%
	without BN		84.88%
DenseNet201	with BN	201	86.02%
	without BN		82.25%
Ensemble Network	–	–	95.63%

Therefore, we changed the loss function of the network, and changed the Cross - entropy loss function to focus loss. This function can make the network pay more attention to the category with fewer samples. Figure 3(b) indicates that compare with cross-entropy loss function CNN model with focal loss function seems to reduce the recognition accuracy of citrus HLB general and citrus HLB. However, from Fig. 3(b), the confusion matrix, indicates that the recognition accuracy of each category has more balance than Fig. 3(a).

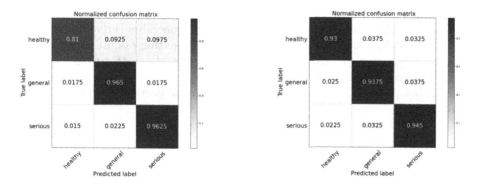

Fig. 3. Confusion matrix Normalized confusion matrix of ResNet50 (Adam, with GAP)

3.2   Experiments on CDy-RELU

The purpose of this experiment is to validate the improved network performance of the proposed activation function, so we contrast the identification accuracy of four single networks in the CDy-RELU using ReLU and. As shown in Table 4, with CDy-RELU, the four individual networks showed different degrees of improvements in recognition accuracy.

Table 4. Performance comparison of ReLU and CDy-ReLU, focal loss function and cross-entropy loss function

Model	Test Accuracy
AlexNet (ReLU, focal loss)	83.28%
AlexNet (CDy-ReLU, focal loss)	84.76%
VggNet19 (ReLU, focal loss)	85.62%
VggNet19 (CDy-ReLU, focal loss)	87.17%
ResNet50 (ReLU, focal loss)	91.25%
ResNet50 (CDy-ReLU, focal loss)	93.58%
DenseNet201 (ReLU, focal loss)	86.02%
DenseNet201 (CDy-ReLU, focal loss)	90.13%

3.3   Compare Single Network and Ensemble Network

To demonstrate that Ensemble Network has higher precision than a single network, in this experiment we tested each CNN model with CDy-ReLU and focal loss. Compare EnsembleNet, As we can observed from Table 5, the EnsembleNet that combines four CNN models have higher recognition accuracy. Of course, compared with Lightweight CNN models such as SqueezeNet [11] and MobileNet [9], EnsembleNet is not perfect. It needs more computing resources for training than a single network, although its performance is also better than that of a single network. We believe that this model approach is more appropriate for computing resource-rich environments than simply porting the model to the mobile side.

Table 5. Compare the performance of single network and EnsembleNet

Model	Test Accuracy
ResNet50	90.35%
VggNet16	93.41%
ResNet50	93.58%
DenseNet	92.76%
SqueezeNet	92.31%
MobileNet	91.49%
Ensemble Network	95.63%

4   Conclusion

In this work, a convolution neural network was used to identify diseases from the images of citrus leaves. To enrich the training data and improve the generalization ability and robustness of the network, the existing data were enhanced using AutoAugment methods. The networks used were AlexNet, VggNet, ResNet, and DenseNet. First, we normalized the above four CNN models in batches and compared their performances. The results showed that the convergence rate and performance of batch normalization are better than those without batch normalization. Replaced the fully connected layer with a global pooling layer and the best ResNet accuracy was 91.18%. We then used a new activation function, concatenated dynamic ReLU, and verify its effectiveness. Meanwhile, the imbalance of the original dataset, and replaced cross-entropy loss focal loss function in ResNet50. The accuracy rate of disease identification from citrus leaves reached 93.58%. Finally, we optimize the single network structure with the above strategy, we integrated them and construct an EnsembleNet, which has better performance than a single network, achieved 95.63% accuracy on citrus leaf dataset.

We believe that the method described in this paper can further improve the identification efficiency of citrus disease, thus providing a basis for the identification of other plant leaf diseases.

Acknowledgements. This paper was supported by the National Natural Science Foundation of China Project No. 61863013, and Hanjiang Normal University's Scientific Research Project 2018A08. This paper was also supported by the National Natural Science Foundation of China (No. 92159102), the Natural Science Foundation of Jiangxi Province (No. 20232ACB205001), Support Plan for Talents in Gan Poyang – Academic and Technical Leader Training Project in Major Disciplines (No. 20232BCJ22025).

References

1. Andrea, C.C., Daniel, B.B.M., Misael, J.B.J.: Precise weed and maize classification through convolutional neuronal networks. In: 2017 IEEE Second Ecuador Technical Chapters Meeting (ETCM), pp. 1–6. IEEE (2017)
2. Arsenovic, M., Karanovic, M., Sladojevic, S., Anderla, A., Stefanovic, D.: Solving current limitations of deep learning based approaches for plant disease detection. Symmetry **11**(7), 939 (2019)
3. Brahimi, M., Boukhalfa, K., Moussaoui, A.: Deep learning for tomato diseases: classification and symptoms visualization. Appl. Artif. Intell. **31**(4), 299–315 (2017)
4. Clevert, D.A., Unterthiner, T., Hochreiter, S.: Fast and accurate deep network learning by exponential linear units (ELUs). arXiv preprint arXiv:1511.07289 (2015)
5. Cubuk, E.D., Zoph, B., Mane, D., Vasudevan, V., Le, Q.V.: AutoAugment: learning augmentation strategies from data. In: Proceedings of the IEEE/CVF Conference on Computer Vision and Pattern Recognition, pp. 113–123 (2019)
6. Fuentes, A., Yoon, S., Kim, S.C., Park, D.S.: A robust deep-learning-based detector for real-time tomato plant diseases and pests recognition. Sensors **17**(9), 2022 (2017)
7. He, K., Zhang, X., Ren, S., Sun, J.: Delving deep into rectifiers: surpassing human-level performance on ImageNet classification. In: Proceedings of the IEEE International Conference on Computer Vision, pp. 1026–1034 (2015)
8. He, K., Zhang, X., Ren, S., Sun, J.: Deep residual learning for image recognition. In: Proceedings of the IEEE Conference on Computer Vision and Pattern Recognition, pp. 770–778 (2016)
9. Howard, A., et al.: Searching for MobileNetV3. In: Proceedings of the IEEE/CVF International Conference on Computer Vision, pp. 1314–1324 (2019)
10. Huang, G., Liu, Z., Van Der Maaten, L., Weinberger, K.Q.: Densely connected convolutional networks. In: Proceedings of the IEEE Conference on Computer Vision and Pattern Recognition, pp. 4700–4708 (2017)
11. Iandola, F.N., Han, S., Moskewicz, M.W., Ashraf, K., Dally, W.J., Keutzer, K.: SqueezeNet: AlexNet-level accuracy with 50x fewer parameters and <0.5 mb model size. arXiv preprint arXiv:1602.07360 (2016)
12. Klambauer, G., Unterthiner, T., Mayr, A., Hochreiter, S.: Self-normalizing neural networks. In: Advances in Neural Information Processing Systems, vol. 30 (2017)
13. Krizhevsky, A., Sutskever, I., Hinton, G.E.: ImageNet classification with deep convolutional neural networks. In: Advances in Neural Information Processing Systems, vol. 25 (2012)

14. Lin, T.Y., Goyal, P., Girshick, R., He, K., Dollár, P.: Focal loss for dense object detection. In: Proceedings of the IEEE International Conference on Computer Vision, pp. 2980–2988 (2017)
15. Picon, A., Seitz, M., Alvarez-Gila, A., Mohnke, P., Ortiz-Barredo, A., Echazarra, J.: Crop conditional convolutional neural networks for massive multi-crop plant disease classification over cell phone acquired images taken on real field conditions. Comput. Electron. Agric. **167**, 105093 (2019)
16. Simonyan, K., Zisserman, A.: Very deep convolutional networks for large-scale image recognition. arXiv preprint arXiv:1409.1556 (2014)
17. Sladojevic, S., Arsenovic, M., Anderla, A., Culibrk, D., Stefanovic, D., et al.: Deep neural networks based recognition of plant diseases by leaf image classification. Comput. Intell. Neurosci. **2016** (2016)
18. Tang, Z., Yang, J., Li, Z., Qi, F.: Grape disease image classification based on lightweight convolution neural networks and channelwise attention. Comput. Electron. Agric. **178**, 105735 (2020)

PM2.5 Monitoring and Prediction Based on IOT and RNN Neural Network

Nengke Xie and Bo Li[✉]

School of Electronic Information Engineering, Shanghai DianJi University, Shanghai, China
libo@sdju.edu.cn

Abstract. With the rapid development of social economy and the increasing improvement of people's living standards, the consumption of fossil energy and the use of transportation are increasing day by day. Leading to a large number of PM2.5 and a variety of air pollutants increasingly serious problems. Therefore, it is very important to monitor and predict air quality and pollutant index in real time, and more accurate algorithms and more convenient visual interfaces can make people better understand the air pollution situation. it is of great significance to analyze the real-time changes in the concentration of air pollutants such as PM2.5 in the atmosphere as well as accurately forecast and early warning. This study monitors the contents of various gases in the atmosphere in real time by building an air quality monitoring platform.

The structure and implementation process of the RNN and the neural network optimized by particle swarm optimization algorithm and genetic algorithm are analyzed and introduced. Various basic data of Nanjing city are taken as data samples, and the RNN under the action of the optimization algorithm is iteratively trained, and then the optimization of the two algorithms for the RNN is compared The improved particle swarm optimization algorithm is more advantageous

Keyword: MQTT PLC AD conversion RNN Industrial gateway PSO ModbusRTU

1 Introduction

In recent years, with the rapid development of economy, the continuous promotion of urban industrialization and the acceleration of urbanization, the problem of urban pollution has become increasingly prominent, especially particulate matter (PM) pollution. Therefore, effective prediction of PM2.5 concentrations and intervention measures are important to reduce health risks. At present, PM2.5 concentration prediction models can be divided into three categories: deterministic model, statistical model and machine learning prediction model.

[2] takes Seoul City as an example and introduces RNN algorithm to improve the previous CMAQ prediction model. Unlike typical neural network algorithms, RNN algorithms learn time series information, so the input data set of PM values and meteorological parameters for the RNN model are sorted according to time.

© The Author(s), under exclusive license to Springer Nature Singapore Pte Ltd. 2024
J. Vaidya et al. (Eds.): AIS&P 2023, LNCS 14510, pp. 241–253, 2024.
https://doi.org/10.1007/978-981-99-9788-6_21

RNN [5] was developed to enable the system to operate automatically and intelligently, which uses historical data to predict future indoor PM2.5 concentrations. The RNN architecture consists of an autoencoder and a loop section.

This paper [12] uses Convnet and Dense-based Bidirectional Gated Recurrent Unit to predict PM2.5 value which combined Convnet, Dense and Bi-GRU. The feature in air quality data was extracted from convnets without max-pooling instead another convolutional layer and Bi-GRU with additional Dense could provide a more accuracy result.

In this study, [11] developed a novel deep-learning-based spatiotemporal interpolation model, which includes the bidirectional Long Short-Term Memory (LSTM) Recurrent Neural Network (RNN) as the main ingredient, this model is able to take into account both spatial and temporal hidden influencing factors automatically.

[14] A multiple attention (MAT) mechanism based on multilayer perception, which includes monitoring site attention, time feature attention, andweather attention, was designed to obtain the spatial-temporal and meteorological dependences of PM2.5 and a hybrid deep learning method based on MAT long short-term memory (MAT-LSTM) neural networks is proposed to predict PM2.5 concentration.

This study [3] used daily data for five years in predicting PM2.5 concentrations in eight Korean cities through deep learning models. PM2.5 data of China were collected and used as input variables to solve the dimensionality problem using principal components analysis (PCA), the deep learning models used were RNN, long short-term memory (LSTM), and bidirectional LSTM (BiLSTM).

Compared with traditional stereotypical and statistical models, machine learning models have good generalization ability and can obtain higher prediction accuracy based on the nonlinear relationship between air pollutants and meteorological factors and PM2.5 concentration, such as random forest, support vector, neural network and other models. RNN has strong ability of nonlinear mapping, learning, self-adaptation and fault tolerance, and is widely used in PM2.5 concentration prediction.

2   Related Works

2.1   IoT for Monitoring Air Quality

IoT [8] can be defined as "An open and comprehensive network of intelligent objects that have the capacity to auto-organize, share information, data and resources, reacting and acting in face of situations and changes in the environment".

Several monitoring system [1] have been proposed recently for air pollution monitoring both in homes and the external environment. For example, a low-cost indoor air quality monitoring wireless sensor network system was developed using Arduino, XBee modules, and micro gas sensors in this reference

Similarly, [10] proposed a low-cost pollution monitoring system. The system utilized semiconductor gas sensors with Wi-Fi module to measure concentration of target gases such as CO, CO2, SO2 and NO2. A Raspberry Pi micro-computer was also provided to act as a base station to handle data transmitted from the nodes and act as a webserver for data visualization.

[7] designed an IoT-based indoor air quality monitoring platform, consisting of an air quality-sensing device called "Smart-Air" and a web server. This platform is dependent on IoT and a cloud computing technology to monitor indoor air quality in real time and anywhere. Smart-Air has been developed to efficiently monitor air quality and transmit the data to a web server via LTE in real time.

the researcher [9,13] designs an Indoor Air Quality Monitoring System based on Arduino which could help to raise the human awareness of air quality, the air quality data which uploaded to ThingSpeak, then retrieved by AirQmon, a customized Android application developed by the researcher to monitor the air quality which is installed on the smartphone.

This article [4], survey the rapidly growing research landscape of low-cost sensor technologies for air quality monitoring and their calibration using machine learning techniques, and identify open research challenges and present directions for future research.

This paper [6] deals with the development of a portable electronic device that simultaneously measures toxic gases and suspended particles in real time. It is based on a microcontroller board and low-cost sensors including dust sensor, smoke sensor, carbon dioxide (CO2) sensor, carbon monoxide (CO) sensor, temperature, and humidity sensors.

2.2   Air Quality Prediction Algorithm

Simple neural networks can only deal with discrete inputs alone, that is, the two inputs are not related before and after, but in some tasks, it is necessary to better deal with the sequence information that the input influences and is related to each other. RNN is one of the many artificial neural networks that can describe the relationship between sequence output and previous information.

Figure 1 is the expansion form of a typical RNN structure diagram, where U is the input layer weight matrix, V is the output layer weight matrix, and W is the hidden layer weight matrix. The input of the network is usually a sequence, and each vector in the sequence will enter the same cell body for operation, resulting in a state vector, and the state vector of the previous moment is input into the next cell together with the sequence vector of the current moment, and so on.

The RNN structure is usually used for sequence problems, and its mathematical expression is as follows:

$$\begin{cases} s_t = f(Ux_t + ws_{t-1}) \\ O_t = g(Vs_t) \end{cases} \tag{1}$$

This network receives input x_t at time$t$,the value of the hidden layer is s_t. Equivalent to each time you have a memory of the previous moment, the output

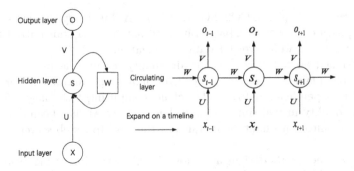

Fig. 1. RNN neural network

value is o_t. It is worth noting that the value of s_t depends not only on x_t, but also on s_{t-1}

The characteristic of RNN is that it can "trace back to the source" and use historical data to solve the "gradient disappearance" is very necessary. The solution to "gradient disappearance" is to choose a better activation function. Generally, the ReLU function is selected as the activation function, and the derivative of the ReLU function in the domain greater than 0 is identical to 1, which can solve the problem of disappearing gradient.

2.3   Neural Network Optimization Algorithm

Particle swarm optimization (PSO) is a population-based stochastic optimization technique, which originates from the study of bird predation behavior. Members of a group constantly change their search patterns through their own experiences and those of other members to find goals in a cooperative manner. suppose there are m particles in a d-dimensional target search space. where the position and velocity of the i-th particle are $x_i = (x_{i1}, x_{i2}, ..., x_{id})$ and $v_i = (v_{i1}, v_{i2}, ...v_{id})$. among $i = 1, 2, ..., m$. for each iteration, the speed and position update formulas are as follows:

$$v_{id}^{k+1} = \omega v_{id}^k + c_1 r_1 \left(p_{id}^k - x_{id}^k \right) + c_2 r_2 \left(p_{gd}^k - x_{id}^k \right)$$

$$(2)$$

$$x_{id}^{k+1} = x_{id}^k + v_{id}^{k+1}$$

$p_i = (p_{i1}, p_{i2}, \cdots, p_{id})$ is the optimal position of the i-th particle currently searched, $p_g = (p_{g1}p_{g2}, ..., p_{gd})$ Is the best location the entire particle swarm has searched so far, $x_{id}{}^k$ and $v_{id}{}^k$ is the position and velocity component of the $i-th$ particle flying in the d-dimension at the $k-th$ iteration. c_1 and c_2 are the learning factors of particles. r_1 and r_2 indicates a random number between $[0,1]$. w expressed inertia weight

Particle swarm optimization algorithm improvement: Inertia weight A affects the performance and efficiency of the algorithm, and appropriate inertia weight can locate the optimal scheme in a small number of iterations. In the traditional particle swarm, the inertia weight is fixed. In this paper, the improved algorithm adopts the variable inertia weight which decreases linearly with the number of iterations. At the early stage of iteration, a larger inertia weight is beneficial to the global search of the algorithm, and at the late stage, a smaller inertia weight can enhance the local search ability of the algorithm. The inertia weight adopted satisfies the following formula $w = w_{max} - k \frac{(w_{max} - w_{min})}{k_{max}}$.

In the formula, w_{max} and w_{min} are the maximum and minimum values of inertia weight respectively, and k_{max} is the maximum number of iterations. Research shows that the performance of PSO can be greatly improved when the inertia weight decreases linearly from 0.9 to 0.4, so the inertia weight in this algorithm can be expressed as the following formula $w = 0.9 - \frac{0.5 \times k}{k_{max}}$.

The optimization process is shown in the following Fig. 2.

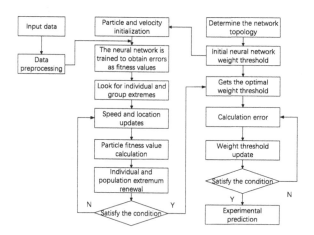

Fig. 2. PSO optimization process

Genetic Algorithm: is a global optimization algorithm inspired by Darwin's theory of natural selection and population genetics. In genetic algorithms, a set of solutions consisting of a set of candidate solutions is called a population. After the initial population is randomly generated, each candidate solution individual is evaluated and selected according to the fitness function, and the individual with high fitness is retained and the individual with low fitness is discarded. After the selection, the combination crossover and mutation operations are carried out on the basis of good individuals to produce new individuals that are more adapted to the environment, that is, the approximate solution that is closer and closer to the optimal solution. (1) The initial population uses real number coding to achieve individual coding, and all individuals are represented by real number strings, including the connection weights of the input layer and the hidden layer, the threshold of the hidden layer, the connection weights of the hidden layer and the output layer, and the threshold of the output layer. The left and right weights and thresholds of the neural network are included in the individuals. Then a neural network with certain structure, weight value and threshold value can be formed.

(2) According to the initial weight and threshold of RNN obtained by individuals, the training prediction output of RNN neural network is carried out by using 80% training data the sum of the absolute squared deviations of the expected and predicted outputs is defined as individual fitness F, the calculation formula is as follows $F = k \sum_{i=1}^{n} \left| y_i{}^2 - o_i{}^2 \right|$, y_i is the expected output of the $i - th$ node of the neural network. o_i is the actual output of the i node, k is coefficient.

(3) Selection operation: select roulette mode for selection operation, and the selection probability p_i of each individual i is as follows $p_i = \frac{f_i}{\sum_{j=1}^{N} f_i}$, $f_i = \frac{k}{F_i}$ and f_i represents the fitness of individual i . N represents the number of individuals in the population.

(4) Cross operation: Multiple individuals are crossed by real numbers, and the cross operation method of chromosome k,l,i in the j position is

$$\begin{cases} a_{kj} = a_{kj}(1 - b) + a_{ij}b \\ a_{lj} = a_{lj}(1 - b) + a_{kj}b \\ a_{ij} = a_{ij}(1 - b) + a_{lj}b \end{cases} \quad ,b \text{ is a random number between } [0,1]$$

(5) Mutation operation: Select the $j - th$ gene a_{ij} of the $i - th$ individual for mutation, and the operation method is as follows

$$a_{ij} = \begin{cases} a_{ij} + (a_{ij} - a_{max})f(g) \, (r > 0.5) \\ a_{ij} + (a_{min} - a_{ij})f(g) \, (r \le 0.5) \end{cases}$$

$f(g) = r_2 \left(1 - \frac{g}{G_{max}}\right)^2$. r_2 represents random number, a_{max} represents the upper bound of gene a_{ij}, a_{min} represents the lower bound of a gene a_{ij}, G_{max} represents the maximum number of evolutions, r indicates a random number between $[0,1]$ .the optimization process is shown in Fig. 3.

3   Methodology

3.1   Monitoring System Design

The air quality prediction system designed in this paper integrates functions such as data acquisition, remote transmission, storage management and remote monitoring and prediction. The overall architecture is shown in the Fig. 4.

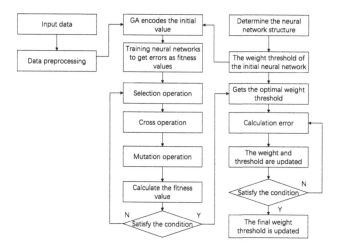

Fig. 3. GA optimization process

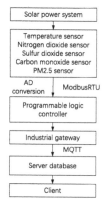

Fig. 4. Air quality prediction system

The data acquisition system uses the solar energy system to achieve self-power supply, the acquisition system uses a variety of industrial sensors for air data acquisition and conversion, and the collected data is transmitted to the programmable logic controller through ModbusRTU communication and AD conversion, and then transmitted to the server through the industrial gateway supporting MQTT protocol. Finally, data monitoring and data analysis are realized in the client software

3.2 Hardware Design

Select the industrial intelligent gateway that supports MQTT protocol to read the programmable logic controller data, the gateway uses high-performance industrial 32-bit communication processor and industrial wireless module, takes embedded real-time operating system as the software support platform, and provides 1 RS232, 5 RS485, 4 Ethernet LAN. Ethernet WAN and WIFI interface, the programmable logic controller adopts Siemens 200smartPLC to connect the analog sensor of data acquisition and communicate with the sensor supporting ModbusRTU as the master station. The analog signal uses a 4–20ma current signal, and the reason for using a current signal is that it is not easy to be disturbed. Current type transmitters convert physical quantities into 4 20mA current outputs. The circuit is shown in the Fig. 5

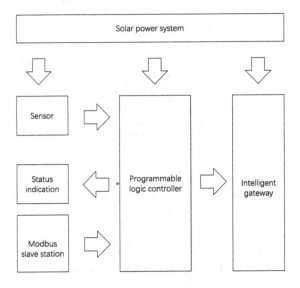

Fig. 5. Hardware system

The power supply of the monitoring system mainly comes from solar energy, and the energy storage device is a battery pack composed of lead-acid batteries. The load of the monitoring system is the main part of the entire system's power consumption, which mainly includes sensor module, programmable logic controller module, intelligent gateway, etc. The sensors of the monitoring system mainly include temperature sensor, nitrogen dioxide sensor, sulfur dioxide sensor, and so on. Carbon monoxide sensor, PM2.5 sensor.

3.3 Monitor Client Platform Design

The monitoring client platform is the core component of the monitoring system. The overall architecture of the network service platform is shown in the Fig. 6

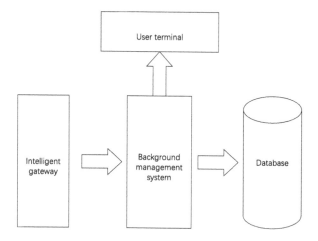

Fig. 6. Network service platform

The server construction mainly includes the end user, database, background information management, etc. The background management system mainly analyzes the MQTT protocol messages to obtain various data, and then stored in the database and transmitted to the interface. The monitoring client platform is mainly developed using C sharp language of visual studio software. The final page looks like Fig. 7

4   Results

Experiment under the Windows operating system, using Python as a programming language, using Tensorflow build neural network prediction model, the software version for Anaconda4.2.0, Python3.5.2, Tensorflow1.3.0. In order to verify the validity of the proposed PM2.5 concentration prediction model, this paper conducts training and testing with the help of relevant data. This paper selects the open environmental data of Nanjing, including the air quality data and meteorological data of Nanjing from June to December 2021 and from June to December 2022.

The data in the Table 1 is the 24-hour average value of Nanjing, and the data unit is ug/m^3. Temperature data is measured in degrees Celsius $°C$, Model evaluation index, this paper predicts the real value of $PM2.5$concentration in the air, aiming at a regression problem. Therefore, root mean square error RMSE, mean absolute error MAE and mean absolute percentage error MAPE were used as three evaluation indicators, respectively as follows:

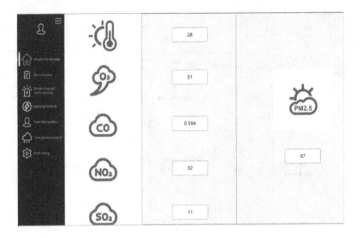

Fig. 7. Monitor client platform

Table 1. Sample data sheet

Date	Maximum temperature	Minimum temperature	NO_2	SO_2	CO	O_3	$PM2.5$
20210601	26	23	32	11	0.594	51	67
20210602	25	20	31	10	0.773	39	56
20210603	28	19	37	15	0.585	99	57
20210604	30	20	53	27	0.74	170	86
20210605	25	18	37	28	0.815	307	90
....	......	......	......	......	......	......	......
20220930	28	19	54	13	0.958	149	51

$$RMSE = \sqrt{\frac{1}{n}\sum_{i=1}^{n}(y_i - y_i^*)}$$

$$MAE = \frac{1}{n}\sum_{i=1}^{n}|y_i - y_i^*|$$

$$\text{MAPE} = \frac{1}{n}\sum_{i=1}^{n}\frac{|y_i - y_i^*|}{y_i}$$

The main parameters of the prediction model are as follows: RNN model with two-layer network structure, activation function as ReLU, and training algorithm as gradient descent. According to the above parameters, a model was established, and the real-time air quality data and meteorological data of the day were used to analyze the PM2.5 of the day in Nanjing. The concentration is predicted. There are 500 groups of data samples, 10% of which are randomly selected as test samples. Some prediction results are shown in Table 2 and Fig. 8.

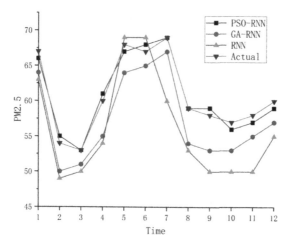

Fig. 8. Prediction

It can be seen from Table 2 that the RNN neural network model optimized by the improved particle swarm optimization algorithm can reduce PM2.5. The concentration is effectively pre-measured, and the error between the predicted value and the real value is small. Due to the small concentration range of PM2.5 in the sample, this paper considers it good that the absolute value of relative error is less than 10%. The total sample is 500, and 10% is selected as the test sample, so the number of test samples is 50. Among them, there are 40 groups with relative error less than 10%, and 5 groups with relative error greater than 20%, accounting for 80%, 10% and 10% of the total test samples respectively. RMSE, MAE and MAPE were used to evaluate the RNN model improved by particle swarm optimization and the RNN model optimized by genetic algorithm. Table 3 lists the performance indicators

Table 2. Relative error table

True value	Predicted value	Error	Relative error
66	62.52	4.473	0.066
41	38.74	1.254	0.031
52	50.45	11.599	0.227
46	49.23	4.804	0.106
111	107.24	7.913	0.071
66	64.75	2.801	0.042
35	34.22	0.561	0.016
...	...	...	...
98	96.42	7.559	0.077

Table 3. Model performance table

Model	RMSE	MAE	MAPE
GA-RNN	8.57	6.77	12.46
RNN	13.95	10.92	25.43
PSO-RNN	4.87	3.43	8.73

5   Conclusion

In this paper, an improved particle swarm optimization algorithm is proposed to optimize the neural network by comparing the improved particle swarm optimization algorithm with the genetic algorithm, so as to establish the PM2.5 concentration prediction model, so as to obtain better generalization ability and faster convergence speed. Accurate prediction of air quality, such as PM2.5 concentration, provides a theoretical basis for controlling the catastrophic impact of air pollution, and has great significance for the integration of meteorological and information disciplines, and has broad application prospects.

References

1. Abraham, S., Li, X.: A cost-effective wireless sensor network system for indoor air quality monitoring applications. Procedia Comput. Sci. **34**, 165–171 (2014)
2. Chang-Hoi, H., et al.: Development of a pm2. 5 prediction model using a recurrent neural network algorithm for the Seoul metropolitan area, republic of Korea. Atmos. Environ. **245**, 118021 (2021)
3. Choi, S.W., Kim, B.H.: Applying PCA to deep learning forecasting models for predicting pm2. 5. Sustainability. **13**(7), 3726 (2021)
4. Concas, F., et al.: Low-cost outdoor air quality monitoring and sensor calibration: a survey and critical analysis. ACM Trans. Sens. Netw. (TOSN) **17**(2), 1–44 (2021)
5. Dai, X., Liu, J., Li, Y.: A recurrent neural network using historical data to predict time series indoor pm2. 5 concentrations for residential buildings. Indoor air **31**(4), 1228–1237 (2021)
6. Jacob, M.T., et al.: Saïdou: low-cost air quality monitoring system design and comparative analysis with a conventional method. Int. J. Energy Environ. Eng. **12**(4), 873–884 (2021)
7. Jo, J., Jo, B., Kim, J., Kim, S., Han, W.: Development of an IoT-based indoor air quality monitoring platform. J. Sens. **2020**, 1–14 (2020)
8. Korade, S., Kotak, V., Durafe, A.: A review paper on internet of things (IoT) and its applications. Int. Res. J. Eng. Technol. **6**(6), 1623–1630 (2019)
9. Nasution, T., Muchtar, M., Simon, A.: Designing an IoT-based air quality monitoring system. In: IOP Conference Series: Materials Science and Engineering, vol. 648, p. 012037. IOP Publishing (2019)
10. Parmar, G., Lakhani, S., Chattopadhyay, M.K.: An IoT based low cost air pollution monitoring system. In: 2017 International Conference on Recent Innovations in Signal processing and Embedded Systems (RISE), pp. 524–528. IEEE (2017)

11. Tong, W., Li, L., Zhou, X., Hamilton, A., Zhang, K.: Deep learning pm 2.5 concentrations with bidirectional LSTM RNN. Air Qual. Atmos. Health. **12**, 411–423 (2019)
12. Wang, B., Kong, W., Zhao, P.: An air quality forecasting model based on improved convnet and RNN. Soft. Comput. **25**(14), 9209–9218 (2021)
13. Waworundeng, J., Limbong, W.H.: AirQMon: indoor air quality monitoring system based on microcontroller, android and IoT. Cogito Smart J. **6**(2), 251–261 (2020)
14. Yuan, H., Xu, G., Lv, T., Ao, X., Zhang, Y.: Pm2. 5 forecast based on a multiple attention long short-term memory (MAT-LSTM) neural networks. Anal. Lett. **54**(6), 935–946 (2021)

An Image Zero Watermark Algorithm Based on DINOv2 and Multiple Cycle Transformation

Xiaosheng Huang and Yi Wu[✉]

East China Jiaotong University, NanChang, China
2021218083500016@ecjtu.edu.cn

Abstract. To address the issue of insufficient robustness in image feature extraction by current zero-watermarking techniques, which negatively impacts performance. This paper proposes an image zero-watermarking algorithm based on DINOv2 and Multiple Cyclic Transformations(MCT). The method first uses a pre-trained DINOv2 model to extract global features from the original image, which have invariance and generalization and can resist various image transformations and attacks. Secondly, it uses MCT to extract robust statistics (mean, variance, autocorrelation coefficient, etc.) from the image's global features and constructs statistical features according to the statistics, which have good stability and anti-interference ability and can enhance the security and robustness of zero-watermarking. Thirdly, it binarizes the statistical features by threshold 0, obtains the image binary features, and combines them with the watermark sequence of the image owner by XOR operation, obtaining the zero watermark. Finally, it generates a zero watermark from the noisy image for copyright verification, which is symmetrical to the process of constructing a zero watermark from the original image. The experimental results show that the proposed zero-watermarking method is robust against different image attacks (such as rotation, scaling, filtering, compression, noise addition, etc.), and its average Bit Error Rate(BER) is 0.17%.

Keywords: Zero watermarking · Dinov2 · Multiple cycle transformation

1 Introduction

With the development and application of digital image technology, digital image is a crucial information carrier, playing an essential role in various fields. For example, in the fields of medicine, education, entertainment, military, etc., digital image is an essential means of transmitting and processing information. According to statistics, more than 3 billion pictures are transmitted globally every day [5]. Digital image also faces various security issues, such as unauthorized copying, tampering, or distribution. Therefore, it is urgent to protect digital images effectively.

© The Author(s), under exclusive license to Springer Nature Singapore Pte Ltd. 2024
J. Vaidya et al. (Eds.): AIS&P 2023, LNCS 14510, pp. 254–263, 2024.
https://doi.org/10.1007/978-981-99-9788-6_22

Digital watermarking technology is an effective means to protect digital image rights [14]. Digital watermarking technology can be divided into two categories according to whether the original image is modified or not: traditional watermarking and zero watermarking. Traditional watermarking is to embed data physically into the host, thus changing the pixel values of the original image [14]; zero-watermarking is a special watermarking technology that does not embed data physically into the host, but only uses its inherent features to construct zero-watermarking signals for ownership verification [9]. It can effectively solve the contradiction between imperceptibility and robustness, and has become a research hotspot in recent years [6]. This paper adopts zero-watermarking technology, because it has better security and reliability than traditional watermarking technology, and will not cause any loss or damage to the original image.

In zero-watermarking technology, extracting robust inherent features of images is the most critical factor to guarantee its performance [6]. Robust inherent features refer to the features that remain unchanged or stable after various transformations or attacks on images, which determine whether zero-watermarking can be accurately detected in noisy images. Many zero-watermarking algorithms based on different feature extraction methods have been proposed. For example, Reference [12] uses the robustness of the mean value of all selected blocks and the size relationship between blocks in the carrier image to construct feature information; Reference [3, 4] transforms the host image into a binary feature image based on visual secret sharing and Sobel edge detection; However, the features based on the spatial domain are sensitive to most image attacks. Reference [14] uses shearlet transform to extract feature vectors from high-frequency subbands, and combines them with watermark information; however, the features based on transformation lack rotation and are not robust enough to geometric attacks. Reference [13] calculates the moments of low-order quaternion generalized polar complex exponential transform, and mixes them with robust features; Reference [7] uses image normalization technology, and obtains Bessel Fourier moments from normalized images for constructing zero-watermarking. However, these zero-watermarking methods use features manually extracted from prior knowledge to construct zero-watermarking, which leads to their lack of generalization ability to resist different image attacks [8]. Reference [6] proposes a method to realize zero-watermarking by using CNN. However, due to the poor optimization of the designed CNN structure, the feature extraction ability is weak, so the robustness of this method is insufficient.

Aiming at these problems, this paper proposes a digital image zero-watermarking algorithm based on the DINOv2 model and MCT. The algorithm aims to use deep learning and signal processing methods to extract robust and stable features of images, construct zero-watermarking, and realize ownership verification. DINOv2 model is a deep neural network model based on self-supervised learning and contrastive learning, which can learn invariant and generalizable image features from unlabeled data [10]. Multiple cyclic transformation is a statistical feature extraction method in signal processing, which can effectively describe the dynamic change and periodicity of signals, and has good stability and anti-interference ability [11]. The main idea and steps of this paper are as follows: firstly, obtain robust global features of images from the last layer transformer output of pre-trained DINOv2 model [10]; secondly, to further enhance the robustness of zero-watermarking, use multiple cyclic transformations to extract statistics (mean, variance, autocorrelation coefficient, etc.) from the global features of images, and construct statistical features according to the statistics; then, binarize the statistical features by threshold 0, obtain the binary features of images, and combine them with the watermark sequence of the image owner by XOR operation, obtaining the zero-watermark; finally, recover the zero-watermark from the noisy image and compare it with the original zero-watermark to realize copyright verification. This process is symmetrical to the process of constructing a zero-watermark from the original image. The innovation points and advantages of this work compared with existing works are mainly reflected in the following aspects:

1. For the first time, this paper applies the DINOv2 model to digital image zero-watermarking technology, using its powerful self-supervised learning and contrastive learning ability to extract invariant and generalizable image features from unlabeled data;

2. For the first time, this paper introduces MCT as a statistical feature extraction method into digital image zero-watermarking technology, using its good stability and anti-interference ability to extract more robust and stable features from global features;

The rest of the article is organized as follows: Section 2 describes the theoretical basis; Section 3 presents the proposed zero watermark algorithm; Section 4 reports the experimental results; Section 5 summarizes this article and prospects for future work.

2   Theoretical Basis

2.1   DINOv2 Model

DINOv2 is a visual feature extraction model based on self-supervised learning, which can learn robust and generalizable features from large-scale unlabeled image data, and can resist some common image processing operations. DINOv2 model adopts Vision Transformer (ViT) as the model architecture [10]. ViT first divides the input image into 14×14 patches, maps each patch to a 768-dimensional vector, and adds a class vector, which serves as the input of the transformer encoder. The transformer encoder consists of multiple self-attention layers and feed-forward layers alternately stacked, which can capture the long-distance dependencies between different positions in the image. Finally, ViT outputs the class vector and performs classification through a linear layer.

DINOv2 model introduces a self-distillation with no labels [2] method based on ViT to achieve the goal of self-supervised learning [10]. The core idea of DINO is to train a student network to imitate the behavior of a more powerful teacher network without any label information. Both the student network and the teacher network are ViT models, but the teacher network has a dynamically updated exponential moving average (EMA) version of parameters. DINO uses a contrastive learning strategy, that is, to maximize the mutual information between local information of the student network and the teacher network, while minimizing the mutual information between global information. This can make the student network learn visual features with invariance and generalization, without being affected by image transformation or noise as illustrated in Fig. 1.

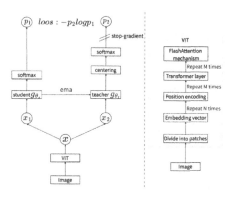

Fig. 1. The structure of DINOv2.

2.2   Multiple Cycle Transformation

MCT [1] is a cyclic transformation technique that can transform a one-dimensional vector into a multidimensional matrix to improve the locality and utilization of data.

The basic idea of MCT is that given a one-dimensional vector x of length N, first choose a period length T and several cycles L. Then, for each $l = 0, 1, \ldots, L - 1$, the vectors x are taken, grouped by period T^l, and the mean of each group is calculated to obtain a vector X_l of length T. Put all X_l together to get a matrix X, which is the result of the MCT. This can be expressed by (1).

$$X_{l,t} = \frac{1}{T} \sum_{k=0}^{T^l - 1} x_{t+kT^l} \quad l = 0, 1, \cdots, L - 1, t = 0, 1, \cdots, T - 1 \qquad (1)$$

MCT has two characteristics: multiplicity and circularity. Multiplicity means that it converts a one-dimensional vector into a multidimensional matrix, and each row represents a frequency component, thus extracting information about different frequencies in the vector. Circularity means that it groups and averages the vector by changing the period length, to make use of the periodic characteristics in the vector to improve the locality of the data and the cache hit rate.

MCT is used to improve the locality and utilization of global image features, from which the statistics are calculated and used to improve the stability and resistance to interference.

3   Zero-Watermarking Algorithm

To improve the robustness of watermarking, this paper proposes a zero-watermarking algorithm based on DINOv2 model. The algorithm uses the DINOv2 model to extract global features from images, and performs MCT on them, obtaining statistical features. Then, it generates zero watermarking according to these statistical features. The same process is applied to the noisy image in the image verification stage. In both zero-watermarking generation and verification stages, the pytorch framework is used to load and use the DINOv2 model's vit-b/14 model to extract global features of images. Next, we will introduce the generation and verification stages of the zero-watermarking algorithm in detail, as illustrated in Fig. 2.

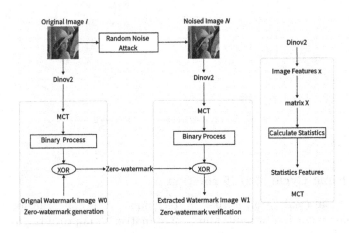

Fig. 2. Zero-watermarking framework based on DINOv2 and MCT.

3.1   Zero-Watermark Generation Stage

The zero-watermark generation stage generates an M_s containing the watermark information using the pre-trained DINOv2 model and the raw image. The specific steps are described as follows:

Step 1. To meet the model requirements and ensure the image quality, resize the image by using the double triple interpolation method to the size of 224×224.

Step 2. Enter the processed image into the pre-trained DINOv2 model, and obtain the global features x of the image from the output of the last layer Transformer of the model.

Step 3. Perform MCT to obtain new features f. The specific steps are described as follows:

Step a. Define the parameters of MCT. They are determined experimentally. When the parameters of MCT are $T = 8$ and $L = 96$, the features obtained by the MCT are more robust.

Step b. Then, perform an MCT on l to obtain an $L \times T$ matrix X, where each row corresponds to a frequency component. Specifically, the elements of the $l - th$ row are calculated by (1).

Step c. Calculate statistics for each frequency component, such as mean, variance, autocorrelation coefficient, etc. The mean vector μ has a length of L, and its elements are given by (2).

$$\mu_l = \frac{1}{T} \sum_{t=0}^{T-1} X_{l.t}, l = 0, 1, \cdots, L - 1 \tag{2}$$

The variance vector σ^2 has a length of L, and its elements are shown by (3).

$$\sigma_l^2 = \frac{1}{T} \sum_{t=0}^{T-1} (X_{l,t} - \mu_l)^2, l = 0, 1, \cdots, L - 1 \tag{3}$$

The variance vector ρ has a length of L, and its elements are shown by (4).

$$\rho_l = \frac{\sum_{t=0}^{T-1} (X_{l,t} - \mu_l)(X_{l,t+1} - \mu_l)}{\sum_{t=0}^{T-1} (X_{l,t} - \mu_l)^2}, l = 0, 1, \cdots, L - 1 \tag{4}$$

Step d. Concatenate the statistics into a feature f, with a length of $3L$, and its elements are shown by (5).

$$f_i = \begin{cases} \mu_i, & i = 0, 1, \cdots, L - 1 \\ \sigma_{i-L}^2, & i = L, L + 1, \cdots, 2L - 1 \\ \rho_{i-2L}, & i = 2L, 2L + 1, \cdots, 3L - 1 \end{cases} \tag{5}$$

Step 4. Binarize the new feature f according to the threshold 0, and obtain a binary feature sequence of length $3L$. That is shown by (6).

$$b_i = \begin{cases} 0, if & f_i < 0 \\ 1, if & f_i \geq 0 \end{cases}, i = 1, 2 \cdots 3L \tag{6}$$

Step 5. Use the binary feature sequence b and the watermark sequence W of the image owner to generate a main share M_s containing watermark information (b, W and M_s have the same shape), and its elements are obtained by (7).

$$M_s = W \otimes b \tag{7}$$

3.2   Image Verification Stage

The image validation phase is the process of extracting zero watermarks from images that may be attacked and verifying image ownership. The specific steps are described as follows:

Step 1. Preprocess the validation images and extract the global features x' with the DINOv2 model.

Step 2. Perform MCT on the features x' to obtain new features f'.

Step 3. Binarize the new feature f' to obtain a binary feature sequence b' of a length of $3L$. This operation is the same as that used in the master share generation phase by (8).

$$b'_i = \begin{cases} 0, if & f_i' < 0 \\ 1, if & f_i' \geq 0 \end{cases} \quad i = 1, 2 \cdots 3L \tag{8}$$

Step 4. Use a binary feature sequence b' and a master sharing M_s to get a recovered watermark sequence W_1. The sequence is obtained by (9).

$$W_1 = M_s \otimes b' \tag{9}$$

Step 5. Verify the image by judging whether the extracted watermark W_1 is similar to W.

4   Experimental Results

To test the effectiveness of the new zero watermarking technique, we generated several distorted images that were not used during the training phase. To evaluate the robustness of the watermark against various attacks, we considered distortions such as JPEG compression with different quality factors, mean filter, Gaussian smoothing filter, and noise contamination.

For the evaluation of the fidelity of the recovered watermark sequence W_1 to its original version W, we used the BER metric. The formula for this metric is provided in (10).

$$BER = \frac{1}{N} \sum_{i=1}^{N} |W_1(i) - W(i)| \tag{10}$$

where, N is the length of the watermark sequence, $W_1(i)$ and $W(i)$ represent the $i-th$ element in the recovered and the original watermark sequence, respectively. Tables 1, 2, 3 and 4 show the experimental results under different attacks and comparisons with other schemes. It can be seen from the table that the proposed algorithm can maintain a low BER under various attacks, indicating that the watermark has strong robustness. In particular, under rotational attacks, the proposed scheme is still able to efficiently recover the watermark, which is difficult for other zero-watermark algorithms.

Table 1. Robustness of watermarks to compression

Quality Factor	PSNR(DB)	BER
10	32.45	0.34%
20	32.98	0.34%
30	34.30	0.34%
40	35.14	0.34%
50	35.81	0.34%
60	36.46	0.34%
70	37.32	0.00%

Table 2. Robustness of watermarking to filtering

Filtering	PSNR(DB)	BER
Mean3×3	32.64	0.34%
Mean5×5	29.26	0.34%
Mean7×7	27.44	0.34%
Mean9×9	26.22	0.34%
Gaussian$\sigma = 1$	32.39	0.34%
Gaussian$\sigma = 2$	28.22	0.34%
Gaussian$\sigma = 3$	26.25	0.34%
Gaussian$\sigma = 4$	24.96	0.34%

Table 3. Robustness of watermarking to Gaussian noise pollution

variance	PSNR(DB)	BER
0.01	15.21	0.00%
0.02	12.91	1.04%
0.005	17.84	0.00%
0.001	25.27	0.00%
0.002	21.91	0.00%

Table 4. Robustness of watermarking to rotation

Angle	PSNR(DB)	BER
1°	29.39	0.00%
2°	28.67	0.00%
3°	28.33	0.34%
4°	28.19	0.34%
5°	28.04	0.34%

5   Conclusion

This paper proposes a zero-watermarking scheme based on the DINOv2 model, which uses the DINOv2 model to extract image features and embed them as watermark information into the main share. In the image verification stage, the image features are extracted again by using the DINOv2 model and compared with the main share, thus verifying the ownership of the image.

The scheme has the following advantages: (1) It does not modify the original image, preserving the image quality; (2) It uses the powerful feature extraction ability of the DINOv2 model to improve the robustness and security of watermarking; (3) It uses the good stability and anti-interference ability of MCT to further improve the robustness and security of watermarking.

The main contribution is to apply the DINOv2 model to the field of zero-watermarking and propose a novel zero-watermarking scheme, which is verified by experiments for its effectiveness and superiority. The limitation is that it does not conduct systematic tests and analysis on different types and sizes of images, nor does it conduct experiments on combined attacks on images. This paper implies that deep learning models can provide more possibilities and potentials for zero-watermarking, which are worth further exploration and research.

The future research directions include: (1) Exploring more deep learning models, such as autoencoders, generative adversarial networks, etc., as the basis of zero-watermarking schemes; (2) Considering more distortion or attack scenarios, such as cropping, scaling, rotating, etc., to improve the robustness of zero-watermarking schemes; (3) Designing more evaluation metrics and standards, such as perceptual quality, security, capacity, etc., to evaluate the performance of zero-watermarking schemes.

Acknowledgements. The authors wish to thank the National Natural Science Foundation of China under Grant 61763011 and the Science and Technology Project of the Educational Department of Jiangxi Province under Grant GJJ00638 and thank the authors of the literature that we use in this article.

References

1. Blahut, R.: Fast Algorithms for Signal Processing. Cambridge University Press (2010). https://doi.org/10.1017/CBO9780511760921
2. Caron, M., Touvron, H., Misra, I., Jégou, H., Mairal, J., Bojanowski, P., et al.: Emerging properties in self-supervised vision transformers. In: 2021 IEEE/CVF International Conference on Computer Vision (ICCV) (2021). https://doi.org/10.1109/iccv48922.2021.00951
3. Chang, C.C., Chuang, J.C.: An image intellectual property protection scheme for gray-level images using visual secret sharing strategy. Pattern Recogn. **23**(8), 931–941 (2002). https://doi.org/10.1016/S0167-8655(02)00008-9
4. Chang, C.C., Lin, P.Y.: Adaptive watermark mechanism for rightful ownership protection. J. Syst. Softw. **81**(7), 1118–1129 (2008). https://doi.org/10.1016/j.jss.2007.11.717
5. Diyi, C.: Global daily transmission of 3 billion pictures, driving the economy and hidden crisis (2021). https://www.yicai.com/news/101076054.html
6. Fierro-Radilla, A., Nakano-Miyatake, M., Cedillo-Hernandez, M., Cleofas-Sanchez, L., Perez-Meana, H.: A robust image zero-watermarking using convolutional neural networks. In: 2019 7th International Workshop on Biometrics and Forensics (IWBF), pp. 1–5 (2019). https://doi.org/10.1109/IWBF.2019.8739245
7. Gao, G., Jiang, G.: Bessel-Fourier moment-based robust image zero-watermarking. Multimed. Tools App. **74**(3), 841–858 (2015). https://doi.org/10.1007/s11042-013-1701-8
8. He, L., He, Z., Luo, T., Song, Y.: Shrinkage and redundant feature elimination network-based robust image zero-watermarking. Symmetry **15**(5), 964 (2023). https://doi.org/10.3390/sym15050964
9. Lang, J., Ma, C.: Novel zero-watermarking method using the compressed sensing significant feature. Multimed. Tools App. **82**(23), 4551–4567 (2023). https://doi.org/10.1007/s11042-019-08406-0
10. Oquab, M., Darcet, T., Moutakanni, T., Vo, H., Szafraniec, M., Khalidov, V., et al.: Dinov2: Learning robust visual features without supervision (2023)

11. Wang, X., Wen, M., Tan, X., et al.: A novel zero-watermarking algorithm based on robust statistical features for natural images. The Visual Comput. **38**(23), 3175–3188 (2022)
12. Xiong, X.: A robust zero-watermarking scheme in the spatial domain **44**, 160–175 (2018). https://doi.org/10.16383/j.aas.2018.c170313
13. Yang, H., Qi, S., Niu, P., Wang, X.: Color image zero-watermarking based on fast quaternion generic polar complex exponential transform. Signal Process. Image Commun. **82**, 115747 (2020). https://doi.org/10.1016/j.image.2019.115747
14. Zhao, J., Xu, W., Zhang, S., Fan, S.X., Zhang, W.: A strong robust zero-watermarking scheme based on Shearlets' high ability for capturing directional features. Math. Probl. Eng. **2016**, 1–11 (2016). https://doi.org/10.1155/2016/4138947

An Image Copyright Authentication Model Based on Blockchain and Digital Watermark

Xiaosheng Huang and Yi Wu[✉]

East China Jiaotong University, Nanchang, China
2021218083500016@ecjtu.edu.cn

Abstract. To solve the problems of digital image leakage and tampering as well as the low imperceptibility of watermarking algorithms in the traditional digital image copyright authentication process, we propose a model that combines digital watermarking, blockchain, and the InterPlanetary File System (IPFS). This model builds an IPFS and blockchain-distributed storage architecture and combines encryption technology to prevent digital images from being leaked and tampered with. Using the immutability of IPFS, we design a watermarking algorithm with high imperceptibility. Our experiments show that the proposed digital image copyright authentication model can effectively prevent digital images from being leaked and tampered with; the mean values of peak signal-to-noise ratio (PSNR) and structural similarity (SSIM) of the proposed watermarking algorithm are 65.607(dB) and 0.999, respectively.

Keywords: Watermarking · Blockchain · IPFS

1 Introduction

Digital images are a type of multimedia data that are widely used in various fields, such as education, medicine, entertainment, military, etc. With the rapid development of technology, digital images can be distributed in digital form through the Internet at any time and place. However, this also makes digital images face serious security risks, such as illegal downloading, modification, copying, and other unauthorized actions. These behaviors not only infringe on the intellectual property rights of the copyright holders but also hinder the progress and application of digital image technology. Therefore, there is an urgent need to effectively authenticate and protect the copyright of digital images.

Digital watermarking technology is a technology that embeds watermark information into digital images to achieve purposes such as copyright protection, identity authentication, data hiding, etc. It uses the human eye's insensitivity to image details to covertly embed watermark information into images, thus realizing the marking and verification of image data. However, this technology also has limitations, such as watermark information being vulnerable to attacks or deletion, The watermark algorithm needs to consider the balance between

J. Vaidya et al. (Eds.): AIS&P 2023, LNCS 14510, pp. 264–275, 2024.
https://doi.org/10.1007/978-981-99-9788-6_23

robustness and imperceptibility [3]. In addition, watermark information requires a neutral third party for storage [9], which also poses problems such as high cost, trust risk, and low efficiency [11]. To promote the progress and application of digital image technology, reducing the cost and improving the efficiency and security of image copyright authentication are important prerequisites for achieving this goal. However relying solely on digital watermarking technology makes it difficult to cope with these challenges, and it is urgent to introduce new technologies and research methods.

Blockchain technology is a distributed database that uses cryptography and consensus mechanisms, which can realize data transparency, tamper-proof, and traceability, thus solving network trust and security problems. Blockchain technology can serve as a distributed third party, storing and providing watermark information, avoiding the problems of a single point of failure, data loss, data leakage, etc. that exist in traditional centralized storage methods [5]. In recent years, there has been an increasing number of studies on the application of blockchain technology to digital image copyright authentication [1, 4, 6, 11]. However, these studies mainly focus on the protection of watermark information, while neglecting the issue of digital image secure storage, which will lead to the risk of digital images being leaked and tampered with.

IPFS technology is a content-addressable distributed file system that can achieve data decentralization, efficient transmission, and permanent preservation, thus improving data availability and reliability [12]. IPFS technology can store digital images in a dispersed and secure network, and access them by unique hash values. There have also been some exploration and attempts at using IPFS technology for digital image storage. Reference [5] effectively prevented digital images from being tampered with by storing them on IPFS. However, this scheme did not make full use of IPFS's tamper-proof property to design more efficient and covert watermarking algorithms; and once the hash value is exposed, it will also pose the risk of digital image leakage.

To address the above problems, we propose a digital image copyright authentication model that integrates digital watermarking, blockchain, and IPFS. First, we build a distributed storage architecture that combines the tamper-proof characteristics of blockchain and IPFS, avoiding the risk of digital images being tampered with; and we use encryption technology to make illegal access visible but unreadable, even if the unique hash value is exposed, the digital image will not be leaked. Next, due to the tamper-proof property of IPFS, the digital images stored on IPFS are not subject to attacks or tampering, so we do not need to consider the robustness of the algorithm when we design the watermarking algorithm, which greatly improves the imperceptibility of the watermarking algorithm. To verify the proposed model and algorithm, PSNR, and SSIM are used as evaluation indicators to evaluate the imperceptibility of the watermarking algorithm.

The structure of this paper is as follows: Section 2 introduces the concepts and principles of blockchain technology and IPFS technology; Section 3 describes the model design and implementation of this paper; Section 4 introduces the watermarking algorithm; Section 5 shows the experiments and analysis of this paper; Section 6 summarizes the contributions and shortcomings of this paper, and puts forward the future research directions.

2   Theoretical Basis

2.1   Blockchain

Blockchain is a new type of database software, first proposed by Satoshi Nakamoto in 2008 [8]. It consists of a data layer, network layer, consensus layer, incentive layer, contract layer, and application layer, as shown in Fig. 1. Blockchain has the characteristics of decentralization, openness, independence, security, and anonymity, which can effectively prevent data tampering and protect privacy. The core of blockchain is that each node stores and verifies all transaction data independently, without relying on any third-party institutions. Blockchain consists of blocks connected in a chain, each block contains a block header and a block body. The block header contains the version number, parent block hash value, Merkle root value, timestamp value, difficulty target value, and nonce value of the block. The parent block hash value is the hash value of the previous block in the chain, which is used to connect the blockchain. The timestamp value indicates the time order of the blocks. The block body consists of the packaged transactions. The hash algorithm ensures that any modification of the transaction changes the hash value, thereby destroying the integrity of the blockchain. Blockchain uses parent block hash value, timestamp, and transaction hash value as technologies to achieve data immutability, which can replace third parties to provide trust services.

2.2   IPFS

IPFS is a peer-to-peer distributed file system that uses content addressing, distributed hash table, BitTorrent protocol, and other technologies, aiming to create a global, permanent, decentralized network storage and sharing system. IPFS works by dividing files into content-addressed blocks and identifying them with unique hash values. When a file is needed, IPFS looks up and assembles all the blocks of that file by hash value. Since IPFS is decentralized, files can be stored on any node and accessed and transmitted efficiently through distributed hash tables. Thus, IPFS ensures the integrity and accessibility of files, while improving the speed and reliability of file transmission. IPFS has advantages such as decentralization, permanence, efficiency, and security. IPFS uses various optimization algorithms to improve the speed of file lookup and transmission, as well as encryption technology and digital signature to ensure the security and integrity of files. IPFS has found wide applications in many fields, such as distributed version control systems, blockchain technology, file sharing and storage, website hosting, etc. For example, Git can store code repositories on IPFS and use IPFS's content-addressing feature to achieve fast code synchronization and

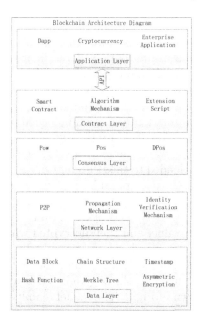

Fig. 1. Blockchain architecture diagram.

merge. Blockchain technology can store large data or metadata on IPFS and verify data integrity by hash values. Users can upload, download, share, and backup various types of files on IPFS and protect their privacy with encryption technology. Website hosting can adopt IPFS as an alternative to traditional servers to provide higher availability, lower cost, and better security. In short, IPFS is a promising technology that provides new possibilities for building decentralized, secure, and efficient network storage and sharing systems.

3   Image Copyright Authentication Model

Based on the demand for digital image secure storage and the characteristics of blockchain and IPFS technologies, we propose a digital image copyright authentication model that combines digital watermarking, IPFS, and blockchain. This model combines digital watermarking, IPFS, and blockchain to achieve digital image protection and verification.

To avoid data duplication on the blockchain and unnecessary copyright checks, we use a perceptual hash function to create a distinctive identifier for each image. Before we initiate the process of copyright registration and verification, we match this identifier with the information already stored on the blockchain. Only when the similarity meets a predetermined threshold, do we execute the corresponding operation. A perceptual hash function differs from a traditional cryptographic hash function, it is robust to regular operations on image content (such as rotation, scaling, cropping) and sensitive to tampering operations on image content (such as deleting, and modifying). A perceptual hash function preprocesses images by reducing size, simplifying color, removing details, and retaining only the structural information of the image. If the structure of the image does not change, its hash value does not change.

The proposed model consists of two main parts: copyright registration and verification. In the copyright registration part, firstly, the watermark is generated according to the watermark information uploaded by the user, and then the watermark information is embedded into the digital image using the watermarking algorithm; secondly, the watermarked digital image and watermark information are encrypted and uploaded to the IPFS network together, and the obtained file address is bound with the perceptual hash value as evidence information and added to the blockchain. In the copyright verification part, first, the encrypted file is downloaded from the IPFS network according to the file address bound with the perceptual hash value, and then decrypted with the key to obtain the watermarked image and watermark information; finally, the watermarked image is detected using the watermarking algorithm, thereby generating a copyright verification report and completing the copyright verification. The schematic diagram of the model is shown in Fig. 2. The model mainly consists of the following steps:

1. The model uses a perceptual hash function to calculate the perceptual hash value of the user-uploaded image
2. Match the perceptual hash value with the information stored in the blockchain, and perform corresponding operations according to the matching results
3. If matched successfully, it means that the image has completed copyright registration, then verify the image
   a. Obtain the encrypted file from IPFS according to the unique hash value bound with the perceptual hash value
   b. Use the key to decrypt the image and perform watermark detection on it, returning a copyright verification report
4. If matched unsuccessfully, it means that the image has not been registered for copyright, then register for image
   a. Embed a watermark in the image and then encrypt it with a watermark
   b. Upload the encrypted image and watermark information to IPFS, add a unique hash value bound with a perceptual hash value to the blockchain
   The model eliminates third-party dependence by using a scheme that combines blockchain and IPFS networks to solve problems such as high cost, trust risk, and low efficiency that exist in traditional third parties; and combines encryption technology to prevent digital images from being leaked and tampered with.

4 Image Watermarking Algorithm

We design an efficient and covert watermarking algorithm that takes full advantage of the tamper-proof property of IPFS. Our watermarking algorithm adopts

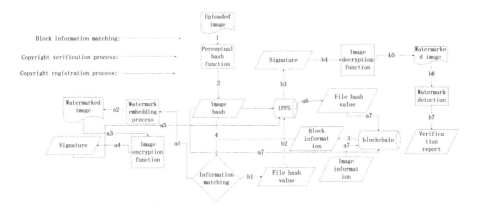

Fig. 2. Model schematic diagram.

the framework in [7], which consists of three steps: watermark preprocessing, watermark embedding, and watermark detection. To adapt to the image watermarking requirements, we use a matrix eigenvalue decomposition model in the watermark preprocessing step, replacing the reversible mapping factor in [7]. Moreover, we also improved the watermark detection step, making it more suitable for image copyright authentication.

4.1   Watermarking Pre-process

Watermark preprocessing is the process of enhancing watermark security by generating a watermark based on the original copyright information. Watermark preprocessing consists of three steps: copyright information preprocessing, matrix eigenvalue decomposition, and position sequence generation. The watermark preprocessing process is illustrated in Fig. 3. The specific steps are as follows:

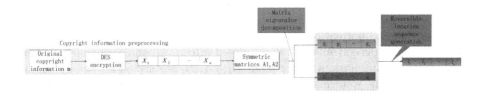

Fig. 3. Watermark pre-process.

Firstly, we encrypt m using key K with DES (Data Encryption Standard) to obtain binary b. We convert b to an n-dimensional matrix A.

$$A = \begin{pmatrix} a_{11} & \cdots & a_{1n} \\ \vdots & \ddots & \vdots \\ a_{n1} & \cdots & a_{nn} \end{pmatrix} \tag{1}$$

The matrix A along the diagonal a_{11} - a_{nn} into two symmetric matrices A_1 and A_2.

$$A1 = \begin{pmatrix} a_{11} & \cdots & a_{1n} \\ \vdots & \ddots & \vdots \\ a_{1n} & \cdots & a_{nn} \end{pmatrix} \quad A2 = \begin{pmatrix} a_{11} & \cdots & a_{n1} \\ \vdots & \ddots & \vdots \\ a_{n1} & \cdots & a_{nn} \end{pmatrix} \tag{2}$$

Then, we perform matrix eigenvalue decomposition on symmetric matrices A_1 and A_2 to obtain the eigenvalue matrices Λ_1 and Λ_2 and the eigenvector matrices C_1 and C_2.we construct two n-dimensional vectors X and Y such that $X = [g_1, g_2, \ldots, g_n]$ and $Y = [g'_1, g'_2, \ldots, g'_n]$.

Finally, we generate the position sequence by using X as the horizontal coordinate and Y as the vertical coordinate, as shown in Table 1.

Table 1. Position sequence generation

X	g_1	g_2	$\cdots$	g_n
Y	g'_1	g'_2	$\cdots$	g'_n
Position	(g_1, g'_1)	(g_2, g'_2)	$\cdots$	(g_n, g'_n)
Position sequence	l_1	l_2	$\cdots$	l_n

4.2 Watermark Embeding

Watermark embedding is the process of embedding the watermark generated by the watermark preprocessing module into the original image imperceptibly according to the position sequence, resulting in a watermarked image that contains watermark information. The steps of watermark embedding are as follows:

First, the copyright image is decomposed into R, G, and B channels. When embedding a watermark, a center point R_1, G_1, B_1 is selected for embedding in each channel. The watermark embedding position is calculated based on (3).

$$\begin{cases} h_i = RGB_c.x + l_i.x \times m_d \\ v_i = RGB_c.y + l_i.y \times m_d \end{cases} \tag{3}$$

Among them, Here,m_d is the dispersion coefficient of the watermark embedding position, which is to prevent the watermark from being excessively concentrated in a particular area. (h_i, v_i) represents the watermark embedding position in the R, G, and B channels.

Next, according to the watermark embedding position (h_i, v_i), the watermark is embedded in the image's R, G, and B channels using (4).

$$RGB(h_i, v_i) = \frac{1}{9} \sum_{k=-n}^{n} RGB(h_i + k, v_i - k) \tag{4}$$

4.3   Watermarking Detection

Watermark detection is the process of detecting a watermark from a watermarked image. The image is then authenticated for copyright based on the results of watermark detection. The steps for watermark detection are as follows:

The original copyright information is first obtained through watermark preprocessing to obtain the watermark position sequence L'. Then, the watermark embedding position is calculated according to (5).

$$\begin{cases} h_i' = RGB_c'.x + l_i'.x \times m_d \\ v_i' = RGB_c'.y + l_i'.y \times m_d \end{cases} \tag{5}$$

Here (h_i', v_i') represents the watermark position coordinate in the watermarked image.

Finally, find the watermark locations in the R, G, and B channels of the watermark image respectively, and determine whether the position contains the watermark based on (6), where ET represents the error range. If the number of watermarks in the image exceeds E, it indicates that the user ID is embedded in the image.

$$\begin{cases} RGB(h_i, v_i) = \frac{1}{9} \sum_{k=-n}^{n} RGB(h_i + k, v_i - k) \\ |RGB'(h_i', v_i') - value_i| \leq ET \end{cases} \tag{6}$$

Table 2. Watermark quality evaluation

Number	Image	PSNR(dB)	SSIM
1	cat	64.190	0.999
2	dog	68.061	0.999
3	wild	64.563	0.999

5   Experiments

To evaluate the imperceptibility of the watermark scheme, we conducted various experiments on the scheme.

In terms of watermark imperceptibility, PSNR is used to objectively evaluate the image quality, which is defined as (7).

$$PSNR = 10\log\frac{\max(I_i(i,j), I_w(i,j))}{\sum\limits_{i=1}^{M}\sum\limits_{j=1}^{N} I_i(i,j) \times I_w(i,j)} \qquad (7)$$

where I_i is the original image, and I_w is the watermark image.

The SSIM value can also be used to evaluate the quality of watermark images. When two images are structurally similar, the maximum value is 1.

Table 3. Compared with blockchain watermarking scheme

Literature	Method	PSNR (DB)	Storage and Distribution	Authentication	Security	Limitations
[2]	Schur decomposition spatial domain watermark	40.000	×	-	-	Security and authentication mechanism limitations
[6]	Rotating vector semi-fragile watermark	41.070	×	Rotating vector watermark	-	Security mecha- nism limitations
[1]	DWT watermark + blockchain	48.147	×	Blockchain	AES watermark encryption + watermark hash	Authentication mechanism limitations
[10]	Blockchain + DCT watermark	39.170	×	Watermark	Arnold transformation	Low watermark imperceptibility
[5]	Watermark + blockchain + perceptual hash function + QR code +v IPFS	-	✓	Blockchain	Perceptual hash function + password hash function	No performance evaluation value
our	Watermark + blockchain + IPFS	65.607	✓	Watermark + image perceptual hash + Blockchain	DES watermark encryption + matrix eigen- value decomposition	-

The structural similarity matrix is given by (8).

$$SSIM(M, N) = \frac{(2\mu_X\mu_Y + C_1)(2\sigma_{XY} + C_2)}{(\mu_X^2 + \mu_Y^2 + C_1)(\sigma_X^2 + \sigma_Y^2 + C_2)} \qquad (8)$$

where, U_x is the mean of I_i, U_y is the mean of I_w, μ_X^2 is the variance of I_i, μ_Y^2 is the variance of I_w, σ_{XY} is the covariance of I_i and I_w. C_1 and C_2 are used to stabilize weak denominator division. L is the dynamic range of pixel values, by default $K1=0.01$ and $K2=0.03$.

Figures 4 and 5 show the watermarked PSNR and SSIM of 500 sizes of 512 × 512 color images with mean values of 64.198 and 0.999, respectively.

In this study, 500 images of cats, dogs, and wild animals were watermarked with 512 × 512 color animal images, and the results are presented in Table 2. The average PSNR was found to be 65.607, and the average SSIM was 0.999.

Table 3 shows the comparison between related papers using blockchain watermarking technology. This comparison is based on 6 categories: methods used during testing, PSNR, copyright work storage and distribution, authentication mechanism, security mechanism, and limitations of work.

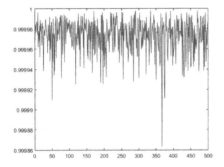
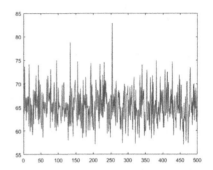

Fig. 4. SSIM line chat.   Fig. 5. PSNR line chat.

6   Conclusion

This paper proposes a digital image copyright authentication model that combines digital watermarking, blockchain, and IPFS, aiming to solve the security risks and copyright issues faced by digital images in network transmission. The main contributions and conclusions are as follows: 1. A distributed storage architecture is constructed, which uses blockchain and IPFS technology to realize the secure storage and access of digital images, avoiding the problems of a single point of failure, data loss, data leakage, etc. that exist in traditional centralized storage methods. 2. An efficient and covert watermarking algorithm is designed, which uses the tamper-proof property of IPFS technology to embed watermark information into digital images, and stores the watermark information and image hash value on the blockchain, realizing the protection and verification of watermark information. 3. Evaluation indicators such as PSNR and SSIM are used to evaluate the imperceptibility of the watermarking algorithm, and comparative experiments are conducted with several other digital image copyright authentication methods based on blockchain or IPFS technology. The results show that the proposed method has higher imperceptibility.

The shortcomings are as follows: 1. The performance and security of blockchain and IPFS technology are not deeply analyzed and tested, and there may be some potential problems and risks. 2. The robustness of the watermarking algorithm under various attack scenarios is not considered, and there may be some possibilities that the watermark information is damaged or tampered with in some special cases. 3. The relationship and influence between encryption technology and watermarking technology are not discussed, and there may be some room for optimization or improvement.

The future research directions include 1. To conduct more in-depth research on the application of blockchain and IPFS technology in digital image copyright authentication, analyze their advantages and limitations, and improve their performance and security. 2. To conduct more diversified and meticulous tests on the watermarking algorithm, consider the robustness of watermark information under various attack scenarios, and improve its resistance ability. 3. To conduct more systematic and theoretical research on the relationship and influence between encryption technology and watermarking technology, and explore more optimized or innovative methods.

Acknowledgements. The authors wish to thank the National Natural Science Foundation of China under Grant 61763011 and the Science and Technology Project of the Educational Department of Jiangxi Province under Grant GJJ00638 and thank the authors of the literature that we use in this article.

References

1. Abrar, A., Abdul, W., Ghouzali, S.: Secure image authentication using watermarking and blockchain. Intell. Autom. Soft Comput. **28**, 577–591 (2021). https://doi.org/10.32604/iasc.2021.016382 https://doi.org/10.32604/iasc.2021.016382

2. Fu, J., Mao, J., Xue, D., et al.: A watermarking scheme based on rotating vector for image content authentication. Soft Comput. **24**, 5755–72 (2020). https://doi.org/10.1007/s00500-019-04318-3

3. He, L., He, Z., Luo, T., Song, Y.: Shrinkage and redundant feature elimination network-based robust image zero-watermarking. Symmetry. **15**(5), 964 (2023). https://doi.org/10.3390/sym15050964, https://www.mdpi.com/2073-8994/15/5/964

4. Li, M., Zeng, L., Zhao, L., Yang, R., An, D., Fan, H.: Blockchain-watermarking for compressive sensed images. IEEE Access **9**, 56457–56467 (2021). https://doi.org/10.1109/ACCESS.2021.3072196

5. Meng, Z., Morizumi, T., Miyata, S., Kinoshita, H.: Design scheme of copyright management system based on digital watermarking and blockchain. In: Proceedings of IEEE 42nd Annual Computer Software and Applications Conference, vol. 2, pp. 359–364 (2018). https://doi.org/10.1109/COMPSAC.2018.10258

6. Su, Q., Yuan, Z., Liu, D.: An approximate Schur decomposition-based spatial domain color image watermarking method. IEEE Access **7**, 4358–4370 (2019). https://doi.org/10.1109/ACCESS.2018.2888857

7. Xiao, L., Huang, W., Xie, Y., Xiao, W., Li, K.C.: A blockchain-based traceable IP copyright protection algorithm. IEEE Access **8**, 49532–49542 (2020). https://doi.org/10.1109/ACCESS.2020.2969990

8. Zhang, E., Li, M., Yiu, S.M., Du, J., Zhu, J.Z., Jin, G.G.: Fair hierarchical secret sharing scheme based on smart contract. Inf. Sci. **546**, 166–176 (2021). https://doi.org/10.1016/j.ins.2020.07.032, https://www.sciencedirect.com/science/article/pii/S0020025520306939

9. Zhang, L., Yang, L., Gao, T.: Image authentication algorithm based on self-embedded robust double watermark. Optoelectron. Laser. **31**(2), 206–213 (2020). https://doi.org/10.16136/j.joel.2020.02.0304

10. Zhaofeng, M., Weihua, H., Hongmin, G.: A new blockchain-based trusted DRM scheme for built-in content protection. J. Image Video Proc. **2018**, 91 (2018). https://doi.org/10.1186/s13640-018-0327-1
11. Zheng, J., Teng, S., Li, P., Ou, W., Zhou, D., Ye, J.: A novel video copyright protection scheme based on blockchain and double watermarking. Secur. Commun. Netw. **2021**(6493306), 6493306 (2021). https://doi.org/10.1155/2021/6493306
12. Zhu, C., Xu, D., Ren, N., Cui, H., Zhao, Y.: A model of geographic data transaction and copyright protection based on blockchain and digital watermarking **12**, 1694–1704 (2021)

Author Index

Printed in the United States
by Baker & Taylor Publisher Services